Rags to Riches: 25 Years of Paper Art from Dieu Donné Papermill

Rags to Riches

25 Years of Paper Art from Dieu Donné Papermill

DONNA STEIN, EXHIBITION CURATOR

MINA TAKAHASHI, EDITOR

WITH A FOREWORD BY CHUCK CLOSE

AND ESSAYS BY SUSAN GOSIN, TRUDY V. HANSEN, AND DONNA STEIN

Dieu Donné Papermill, Inc.
New York, New York

This catalogue is published in conjunction with the touring exhibition
Rags to Riches: 25 Years of Paper Art from Dieu Donné Papermill.
The exhibition itinerary follows:

KRESGE ART MUSEUM
Michigan State University, East Lansing, Michigan, 4 September – 28 October 2001

MARYLAND INSTITUTE COLLEGE OF ART
Baltimore, Maryland, 9 November – 16 December 2001

MARIANNA KISTLER BEACH MUSEUM OF ART
Kansas State University, Manhattan, Kansas, 2 April – 30 June 2002

HECKSCHER MUSEUM OF ART
Huntington, Long Island, New York, 23 November 2002 – 26 January 2003

MILWAUKEE ART MUSEUM
Milwaukee, Wisconsin, 11 April – 22 June 2003

FORT WAYNE MUSEUM OF ART
Fort Wayne, Indiana, 26 July – 2 November 2003

The exhibition and catalogue have been made possible by grants from the New York State Council on the Arts, the National Endowment for the Arts, and The Andy Warhol Foundation for the Visual Arts. Additional financial support has been provided by Fifth Floor Foundation, Furthermore/the publication program of the J.M. Kaplan Fund, Menasha Corporation Foundation, and an anonymous foundation.

Published by Dieu Donné Papermill, Inc.
433 Broome Street, New York, NY 10013
tel 212.226.0573 fax 212.226.6088
info@papermaking.org

ISBN 0-9704072-4-6
Library of Congress Control Number: 2001090524
First Edition
Printed in Canada

Cover photograph: Susan Gosin

Contents

Foreword

In 1981, when Joe Wilfer first took me downtown to Dieu Donné Papermill to introduce me to the idea of translating my "heads" into paper pulp, I was curious but doubtful that this was an appropriate medium for my imagery. Despite my doubts, Joe's tremendous enthusiasm for papermaking's potential and his inner conviction that our collaboration would produce results convinced me to give it a try.

Working from an earlier watercolor of *Keith*, I used a gray scale to reduce the image to a list of numbers corresponding to the relative darkness and lightness of the watercolor. We tinted the pulp with pigment in a related range of grays from white to black and inserted the wet pulp into a plastic grill to produce discrete pads of colored pulp on a wet pulp field. Initial results were disappointing, and I was ready to give up on the process. With Joe's nudging and unrelenting optimism we persevered through many more experiments until finally it began to pay off.

As we worked to adjust values and try various pulps, I began to realize how Joe had seen the possibilities of my imagery – even before I had – in this unfamiliar art medium. It was the start of a particularly enduring collaboration. As I became more used to the process, I realized that I could hand-manipulate the wet pulp before it was pressed to create unexpected variations. I also recycled the unused pulp by using dried chips to construct collage paintings. Throughout all this work together, Joe's style of playful problem solving, remarkable vision, and constant joy in collaboration acted as a catalyst for my evolving imagery. Today, I am once again at Dieu Donné, working with Paul Wong and Ruth Lingen, Joe's successor at Pace, on new self-portraits in stenciled pulp.

Joe and the people at Dieu Donné epitomize the value of collaboration for the visual artist. With team effort, the finished art is greater than the sum of the parts – a product of the distinct mix of people, process, and creative interaction. In a sense, collaboration is chemistry. The people bring their talent, knowledge, and skills, along with a mutual appreciation and trust in each other's work in the studio.

As you will see from the exhibition and this publication, Dieu Donné has been able to provide artists a professional studio and a dedicated and skilled staff who share their enthusiasm and expertise in this unique medium. I invite you to join me in congratulating Dieu Donné on their quarter-century of fine work and look forward to another twenty-five years of successful collaborations. Dieu Donné is a particularly valuable resource for any artist's paper needs from custom sheets to collaborative projects.

CHUCK CLOSE

Acknowledgments

Through this publication, we not only recount the twenty-five year history of Dieu Donné Papermill, but we also mirror a larger story of artistic experimentation which is particularly American. Printmaking, glass arts, ceramics, and the book arts, to name a few, have all experienced a veritable leap from technique-based craft to concept-driven artistic practice, mainly through the pioneering efforts of American artists. Hand papermaking had an early run in America from 1690 through the mid 1800s when the Industrial Revolution arrived and machines completely took over paper production. Other countries, such as Japan, France and Italy, have hand papermills that have been operating uninterrupted for over eight centuries. Whereas the burden of historical applications may deter artists from exploring these long-standing traditions as new forms of expression, freedom from the past enables artists to approach these practices with fresh energy and new ideas. By coupling this forward-looking spirit with a respect for the process and the traditions from which we can learn, "the possibilities are endless," as Trudy V. Hansen reflects in her excellent overview of the developing field of paper art.

We are indebted to guest curator Donna Stein for her summary of Dieu Donné's contribution to papermaking as an artistic practice. She combed through thousands of slides, viewed every work in the Dieu Donné Archive of Paper Art, and made studio visits in order to select work for *Rags to Riches: 25 Years of Paper Art from Dieu Donné Papermill*. By organizing her selections and curatorial essay in sections related to different artistic techniques, Donna provides in-depth descriptions of our studio processes as they are married to the conceptual and working methods of the artists with whom we collaborate. We hope that this information will add to the working knowledge of the field, promote a better understanding of the technical aspects of our work, and inspire artists to discover and visualize how their work can be developed in hand papermaking.

Our work at Dieu Donné represents an inherent partnership between technique and artistic inspiration and, furthermore, a partnership between people and process and among people. Founder and co-chair Susan Gosin shares her perspective on this important aspect of Dieu Donné in her warm and spirited essay describing the founding and development of the organization. Collaboration is further considered in Donna Stein's insightful interview with our artistic director, Paul Wong. As a key member of the Dieu Donné team, Paul has worked with more than three-hundred artists in direct one-on-one collaborations in the studio for twenty-three of Dieu Donné's twenty-five years.

In addition to showcasing Paul's extraordinary talents as a master collaborator and artist, *Rags to Riches* also features important collaborative projects spearheaded by other staff members active on the wet-floor, most notably Pat Almonrode, who was with us at the Mill from 1997 through 2001. When appropriate, we have listed the collaborator and assistants on the checklist for each work in the exhibition.

With *Rags to Riches'* touring schedule extending to nearly three years, Dieu Donné is grateful to all of the lenders and the thirty-nine artists featured in the exhibition, especially our three commissioned artists, Lynda Benglis, Alan Shields, and Michelle Stuart, who created new work for our 25th anniversary. We are pleased that this exhibition will travel well beyond New York to be seen in Michigan, Maryland, Kansas, Wisconsin, and Indiana, and we thank all of the institutions that agreed so enthusiastically to present *Rags to Riches*. Special thanks to Ann Cohen DePietro, chief curator of the Heckscher Museum, for coming up with the exhibition title. I also wish to recognize Sullivan Street Frames for their extraordinary talent as our exhibition framer and Kym Kulp, registrar on this project, who has shared her expertise on the intricate details of traveling exhibitions.

This publication, which accompanies the touring exhibition and serves as our first comprehensive history book, has been a daunting but thoroughly rewarding project. Our warmest thanks to Chuck Close, who opens this publication with his kind and heartfelt foreword. Many thanks again to Trudy V. Hansen, Susan Gosin, and Donna Stein, who each contributed an important essay to this publication, along with Paul Wong, who shared his thoughts in the interview with Donna. I would like to express my appreciation to Donna Stein and Susan Gosin for their guidance and assistance throughout the editorial process.

Our board of directors has also been supportive during this project. I especially thank co-chair James Abruzzo for his guidance, Karen Davidson for her advice, and Michael Wolfensohn for introducing us to photographer Bill Milne, who in turn generously contributed his services to the publication.

I would like to express my deepest appreciation to Dieu Donné staff members Pat Almonrode, John Currie, Stacey Lynch, Megan Moorhouse, and Paul Wong, who shared their time researching, compiling, and proofreading, as well as tapping into their memory banks to help recall the many artists, interns, and volunteers, noteworthy events for our chronology, and published material for our bibliography. It was a real team effort in true Dieu Donné fashion. Even though we have checked and rechecked the manuscript many times over, we know we probably have misspelled someone's name or missed someone very dear and important to Dieu Donné. We apologize in advance to the reader and anyone affected, and ask for your understanding. We welcome and appreciate corrections and other pertinent information for our next history book!

We are grateful to our numerous outside readers for the varied components of this publication including Jacqueline Asplundh, Gail Deery, Helen Hiebert, Julia Miller, and Melissa Potter. We thank the many lenders of photographic material for this publication, especially Mara Leimanis at Pace Editions, Inc., who patiently responded to my numerous requests. We are fortunate to have worked with Sue Henger, who gave the entire manuscript a final read-through and expertly digested and indexed the material presented in this publication.

It has been a pleasure to work again with Russell Maret, who conceived a wonderful design for this publication based on several brainstorming sessions with Susan Gosin and me in which we talked a lot about water, mother earth, and flying over clouds. Russell followed through brilliantly on all the minute details involved in the production end with our fine printers at Friesens.

We would like to acknowledge the following public agencies and foundations for their generous financial support of the *Rags to Riches* exhibition and publication. New York State Council on the Arts provided early and critical support. We are indebted to Elizabeth Merena, NYSCA's director of the visual arts program, for her enduring and invaluable support throughout the years. We are also grateful for the significant support of *Rags to Riches* provided by The Andy Warhol Foundation for the Visual Arts and the National Endowment for the Arts. Other major support was contributed for this project by Fifth Floor Foundation, Furthermore/the publication program of the J.M. Kaplan Fund, Menasha Corporation Foundation, and an anonymous foundation.

The sheer number of artists and constituents who have shared their vision, energies, and passion to help us turn rags into riches is staggering. At the back of this publication we have attempted to list all of the artists who have collaborated with us in the studio, custom-ordered special paper for their projects, contributed work to our exhibitions, or supported us through their generous donation of work to our auctions and archive. We have also listed other clients such as publishers of prints and limited-edition books, conservators, and institutions with whom we've worked, and clients of our educational programs, which have promoted hand papermaking to thousands of people from preschoolers to seniors. In addition, we have noted the funders and organizational lists of dedicated staff, board members, instructors, lecturers, interns, work-exchange students, and volunteers who have helped to implement our many programs, but more importantly, who share Dieu Donné's mission to advance the art of hand papermaking. In compiling all of these lists, along with the chronology and bibliography of the organization, I have had the pleasure of recalling countless experiences that have not only enriched my life but in essence make up the overall history of Dieu Donné. A tremendous number of individuals have contributed to the living story and vitality of Dieu Donné – we are indebted to all of you – thank you.

Finally, on behalf of the staff and the board of directors, I would like to dedicate this publication to Susan Gosin, founder and co-chair of Dieu Donné, for her vision and passion – extraordinary and irresistible – which have shaped and propelled the organization throughout its first twenty-five years and no doubt will for the next quarter-century.

MINA TAKAHASHI

11 June 2001 – Board and Staff of Dieu Donné
from top – against wall: *Brett Littman, Eric Owen Costello, James Abruzzo, Terry Savage, Matthew Jackson, Paul Wong, Cathy Lobel, Karen Davidson, Stephen Yablon.*
against railing: *Janice deYoung, Megan Moorhouse, Laura Stirton Aust, Catherine Bindman, Pat Almonrode, Stacey Lynch, Rachel Fong, Erin O'Rourke, Michael Wolfensohn.*
front row from left to right: *Susan Gosin, Mina Takahashi, Margaret Pomeroy Hunt, Walter Thinnes.*
not pictured: *John Currie, Gail Deery.*

Lenders to the Exhibition

George Adams Gallery, *New York, New York*
Sarah Brayer
Lynda Benglis
Cheim & Read, *New York, New York*
Nancy Cohen
Eyal Danieli
Elena del Rivero
Dieu Donné Papermill, Inc.
Dieu Donné Press, Inc.
Lesley Dill
Ming Fay
Vincent FitzGerald & Co., *New York, New York*
Elizabeth Harris Gallery, *New York, New York*
Virginia Jaramillo
Mary Judge
Nora Shwide LeMorin
Golda Lewis
Lower East Side Printshop, *New York, New York*
Winifred Lutz
Osiris, *New York, New York*
Laurie Ourlicht
Ken Polinskie
Juan Sánchez
Barbara Schwartz
Arlene Shechet
Alan Shields
Sandy Skoglund
Sprint Corporation Art Collection, *Overland Park, Kansas*
Susanna Starr
Michelle Stuart
Mira Takahashi
Bart Wasserman
Paul K. Wong

Pulp and Possibilities: A Brief Look at Contemporary Paper Art

TRUDY V. HANSEN

Throughout the rapidly evolving and expanding art world of twentieth-century America, papermaking – and the use of pulp and paper in artmaking – has engaged artists as diverse as the multifaceted medium itself.[1] Even as modern mid-twentieth-century pioneers of paper art were exploding notions of papermaking primarily as a fine craft, other artists were discovering the thrill of seeking out exquisite papers from around the globe to utilize for their prints, paintings, collages, books, and drawings. Versatile in form and handling, paper and paper pulp have found limitless two- and three-dimensional applications in contemporary art. It is difficult to think of an artist working today who has not in some way incorporated paper into his or her work. Pulp and paper have been the vehicles for some of the most expressive and profound visual statements of our times.

Conceptually, paper has made sense as a medium well suited to contemporary art. With global sources and histories, paper in its many familiar forms could speak to artists and viewers of diverse backgrounds. Unlike the Eurocentric medium of oil paint on canvas, paper was a natural material for artists of wide-ranging cultures and ethnicities who were drawing upon and addressing their own personal visions, artistic traditions, and cultural histories. Just as the carefully tracked and promoted trends based on style and the marketplace broadened to embrace conceptual art, outsider art, installation and performance art, and "low art" as opposed to "high art," paper served as an Everyman's medium. As contemporary Western cultures came to accept Art – the making of art, the forms of art, and the functions of art – as more than the production of precious objects for an elite segment of society, paper became a natural choice for many artists by virtue of its availability, flexibility, and suitability for their projects. Whatever the need, pulp and paper became viable and highly desirable media for an increasingly pluralistic art world.

There is no question that the fortunes of contemporary papermaking and paper art are closely bound to the unprecedented interest in printmaking during the second half of the twentieth century. As large numbers of artists began to experiment with printmaking, a refocused interest in the paper support for their images led artists to demand a greater variety of papers, and that soon led to every manner of exploration with paper pulp itself. As fine papers from Japan, France, England, and other countries were imported in large quantities, interest in global papermaking traditions led both artists and papermakers to turn their attention to different ways of thinking about paper. The simultaneous growth of professional papermaking workshops in the United States soon provided artists with exciting, often custom-made papers for their work as well as a steady supply of high-quality standard papers.

In another fundamental way, however, the intertwined histories of contemporary papermaking and printmaking marked a far more important development. As papermakers began working with artists to create papers for specific projects, a new collaborative arena emerged that matched the energized environment of the contemporary print workshop. As artists were introduced to new paper possibilities, many came to papermills to work in the same way they were collaborating in printshops. Once again (and in spite of the often misguided but persistent lionization of the Abstract Expressionists as lonely creative spirits working in relative isolation), the nineteenth-century romantic notion of the artist as solitary creator was challenged and diminished. Artists, master printers and papermakers, and sometimes dealers and publishers all became part of the process of making art. Many artists sought out opportunities to work at papermills, and they found receptive papermakers ready to add their special talents to projects. Still other artists, having been introduced to papermaking at papermills, workshops, or in university art classes, opted to set up papermaking facilities in their own studios to continue their own paper explorations.

It is important to note that contemporary papermakers, like the heads of print workshops, differed significantly from the guild-based master craftsmen of previous centuries. Many had attended art schools and been trained as painters and sculptors, and they often remained actively engaged in creating their own art or in conducting ongoing research and experimentation into the qualities of paper outside the bounds of specific collaborations. In this setting, the exchange of ideas and skills became a subtle and often continuing dialogue that might last the course of a project or for years. In tandem with printmaking, the field of papermaking served as the vanguard of a new collective attitude and approach to artmaking. Clearly, the collaborative spirit engendered by print and papermaking workshops helped set the stage for other emerging collaborative forms such as performance art, multimedia art, and interdisciplinary projects created by groups of artists, musicians, actors, and writers.

This open and dynamic spirit helped break down barriers between artmaking media and contributed to an explosion of creative exploration and growth on the eve of the Information Age.

Finally, the physical qualities of paper and paper pulp were ideal for vastly different types of artwork. Thick, thin, transparent, opaque, fragile, or as strong as fabric, paper could be torn, sewn, folded, pierced, embossed, and manipulated in countless ways. Paper pulp could be cast, built up in layers, watermarked, sculpted, sprayed, and used as a painting medium for artists to work freely or pour into molds. Depending on the fibers used and the methods of production and finishing, papers could be smooth or heavily textured, rough or refined. Pulp could be dyed, natural or bleached, and infinite materials could be incorporated into the pulp or paper sheets. Paper lent itself to drawing, painting, printing, collage, photography, sculpture, bookmaking, large-scale installations, and assemblage. Few media presented as many possibilities.

While printmaking and an interest in paper have naturally gone hand in hand, interest in paper went far beyond that of a mere support for a printed image. The roots of a broader interest in paper and pulp were firmly established at the turn of the twentieth century as part of the international Arts and Crafts movement. Among the key beliefs of William Morris (1834-1896), the English founder of the movement, were a commitment to hand craftsmanship, the belief in an all-encompassing design ethic that included fine arts, decorative arts, and utilitarian objects, and an appreciation of and respect for materials. The movement in America had many manifestations and practitioners.[2] Beginning in the first decades of the century, the efforts of Dard Hunter – artist, graphic designer, paper historian, and tireless advocate of paper – must be acknowledged.[3] It was Hunter (1883-1966) who became the bridge in papermaking's evolution from craft to art.

Hunter, an early member of Elbert Hubbard's famous Roycroft Studios, was the consummate Arts and Crafts movement artist in America, designing books, decorative objects, and stained glass *(fig. 1)*. After studying papermaking in England in 1911, Hunter returned to the United States and began a lifelong study of paper that culminated in the publication of his definitive history of paper and printing in 1943.[4] He also established a papermill, and although the Lime Rock Mill (1928-31) in Connecticut was not destined to

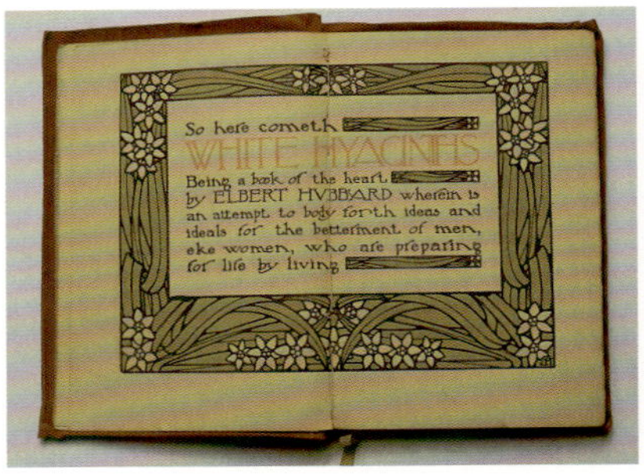

fig. 1 Dard Hunter, Frontispiece for White Hyacinths *by Elbert Hubbard, 1907, 7 1/2 x 10 inches (open)*

become a commercial success, the mill provided inspiration for a host of papermakers later in the century. Artists often consulted Hunter when they were looking for special papers for their work. Hunter traveled throughout the world researching the history of paper and amassed a large library, collection of paper samples, and papermaking artifacts. His collection, the Dard Hunter Paper Museum, was first opened to the public at the Massachusetts Institute of Technology in 1939.[5]

Working in vastly different styles and mediums, many artists during the first half of the twentieth century were intensely interested in paper. Gustave Baumann (1881-1971), for example, found papers for his color woodcuts at German papermills. For later prints, he ordered custom-made papers which were watermarked with his distinctive hand and heart monogram.[6] American *japonistes* such as Helen Hyde (1868-1919) and Arthur Wesley Dow (1857-1922) traveled to Japan to work and study and found a wealth of fine handmade papers available for their prints. Edna Hopkins Boies (1872-1937) experimented with Japanese mulberry papers for her color woodblock prints. In 1903, she wrote in one of her notebooks that "the color was often caught up, as it were, on the outer fibers only, the very whiteness of those below shining through and diluting it with light."[7]

Stanley William Hayter (1901-1988) established print workshops in Paris and New York[8] and encouraged fellow artists to explore paper in their intaglio prints. Hayter himself combined embossing and debossing (also referred to as "inkless intaglio") with sophisticated intaglio printing techniques that resulted in prints of amazing texture and visual complexity. Hayter also kept a supply of fine papers on hand at the atelier; when some artists were not able to buy the expensive papers, Hayter would often encourage them to use it anyway.[9] Hayter's spirit of experimentation, his generosity, and his open workshop that welcomed European and American, young and old, famous and unknown artists to work and share ideas and techniques were important in introducing artists to fine papers and ways to enhance prints through paper manipulations. Artists experimenting with modernist styles also embraced collage, building images with paper both precious and "found." Anne Ryan (1888-1954), for example, made hundreds of small mixed media collages during the 1940s and 50s, using papers ranging from newsprint to handmade papers in combination with bits of fabric, paint, and other materials.

Despite a growing appreciation for paper in the first half of the century, the true "paper revolution" needed another groundbreaking paper protagonist to introduce radically different possibilities for paper. Douglass Morse Howell (1906-1994) was that person.[10] Like Dard Hunter, Howell was artist, paper researcher and scholar, teacher, collaborator, and advocate. Howell's early introduction to fine paper, seen in European art collections and libraries he visited as a youth, instilled a desire to research plant fibers and paper history and to experiment with papermaking techniques. Howell established a papermill in 1946, and his early handmade papers were used by artists including Jackson Pollock, Dorothy Dehner, Chaim Gross, Lee Krasner, Elaine de Kooning, Alfonso Ossorio, Karl Schrag, Jasper Johns, and many others.

Douglass Howell was also creating his own work *(fig. 2)*, beautiful free-form sheets of textured paper, sinuous paper constructions, and intricate paperworks incorporating linen cords and other natural materials that emphasized the expressive and aesthetic qualities inherent in the medium. The primacy of Abstract Expressionism during this period, a movement that permitted gesture and abstraction to replace pictorial subject matter, freed artists like Howell to concentrate on process and materials. When Howell's "papetries" were first exhibited at the Betty Parsons Gallery in 1955, the new genre of paperworks was born. This was also an important year for book arts, as artist Claire Van Vliet (b. 1933) founded Janus Press. Van Vliet's books and paperworks and her extraordinary collaborations with writers, poets, printmakers, and other papermakers helped promote a surge of interest in artists' books.[11]

Of key importance to the paper movement in the 1950s and 60s were printers and print publishers, most notably June Wayne and Tatyana Grosman, who welcomed papermakers into the contemporary printmaking scene. Wayne made an ongoing plea for new American sources of handmade paper as part of her lifelong advocacy in support of American printmaking.[12] In 1959, Howell worked with

fig. 2 Douglass Morse Howell, America, *early 1960s, 17 x 22 inches*

Wayne to make large sheets of handmade paper adhered to canvas for her paintings.[13] Grosman, who firmly believed that the paper was an integral part of the artwork, felt that the paper itself could be a great source of inspiration to the artist.[14] It was Grosman who encouraged artists such as Jasper Johns and Larry Rivers to use the handmade papers Howell was producing and who introduced Howell to artists at her Universal Limited Art Editions on Long Island (also the site of Howell's papermill). One project destined to become an icon of contemporary art was *Stones*, a portfolio of lithographs and poems created between 1957 and 1959 by Rivers and Frank O'Hara and printed on paper that Howell made from old blue jeans. Artists and printers alike were intrigued, and they came to Howell to experiment and collaborate as well as acquire paper. Howell's first collaborations with Michael Ponce de Leon (b. 1922) in 1962 resulted in thick and sturdy long-fibered papers that could withstand the tremendous pressure of a hydraulic printing press. Howell's collaborations, his workshops, his writings, and his art touched countless artists, each coming away with a new appreciation of the medium.

Thus, the stage was set by the 1960s for a wave of innovation in the field of paper art. Many factors contributed to acceptance of the evolving genre. With the shift of the international art scene from Paris to New York after World War II, a new interest in contemporary art arose in the United States. Galleries opened, museums began to collect and exhibit contemporary work, and there was greater coverage of the art scene in newspapers and periodicals. The postwar economic boom, and an influx of returning servicemen who were able to take advantage of the G.I. Bill to attend college, led to the rapid growth of college and university art departments. Growing affluence led to a growing art market, and increased demand for artworks from a young, art-savvy class of collectors helped spur an interest in works on paper. Well-known artists were encouraged by their dealers to make prints and paperworks in addition to their large-scale paintings and sculpture, and many welcomed the opportunities to explore new mediums and processes. During an era that worshipped the new, this 2,000-year-old medium was suddenly new and sexy.

Papermills, like print workshops, began to spring up across the country. Douglass Howell directly inspired many papermakers. Laurence Barker *(fig. 3)*, who had established the first college-level courses in papermaking at Cranbrook Academy in Michigan, had studied with Howell at his papermill in Long Island, as had Golda Lewis, who in the early 1960s was one of the first women to work seriously with paper.[15] Barker's students went on to set up shops, both independently and in university settings. Walter Hamady went to the University of Wisconsin and established a papermill, The Shadwell Papermill, and The Perishable Press, which was devoted to the production of handmade books *(fig. 4)*. Aris Koutroulis went to Wayne State University, and John Koller, along with his wife Kathleen, established HMP (Hand Made Papers) in Woodstock, Connecticut, in 1973. Winifred Lutz *(fig. 5)* went to teach at Yale and in the sculpture department at the Tyler School of Art in Philadelphia, establishing a strong career that included groundbreaking multidimensional paperworks and installations. It is a career that continues to this day and, like her predecessors, Lutz has had a tremendous impact on a new generation of artists working with paper.

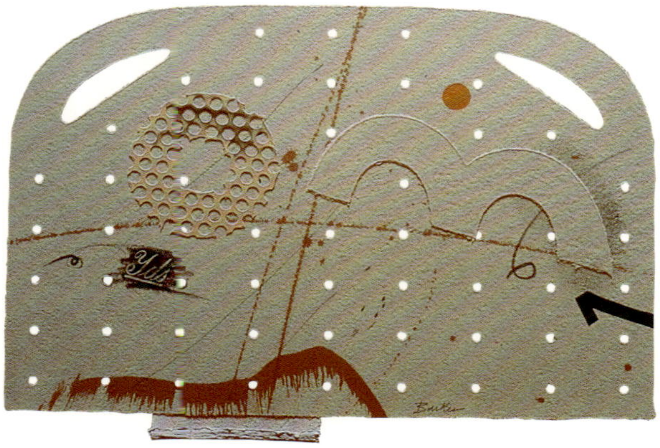

fig. 3 Laurence Barker, Untitled, *1979, 17 x 25 3/4 inches*

The circle continued to widen throughout the 1970s. The Upper U.S. Papermill, started by Joseph Wilfer in Oregon, Wisconsin, in 1969; Twinrocker Papers, established by Kathryn and Howard Clark and Kathryn's twin sister Peggy in 1971 and located in Brookston, Indiana; Farnsworth and Company Papermakers (later Magnolia Editions), started by Donald Farnsworth in San Francisco in 1973 and later moving to Oakland, where he was joined by David Kimball; Dieu Donné Press and Paper, founded in 1976; and Exeter Press, started by Lynn Forgach in Cleveland Heights, Ohio, in 1977, were only a few of the papermills making fine papers and collaborating with artists.[16] Each shop had its own unique personality, some producing custom and stock lines of standard papers, some collaborating with artists, some specializing in paper for book projects, some teaching papermaking and book arts, and many doing several or all of the above. Numerous printmaking workshops also began to develop papermaking facilities in response to the interest shown by their artists in exploring paper. Kenneth Tyler, whose reputation for utilizing innovative technologies and creatively exploring printmaking materials is legendary, was also drawn to experiment with paper after meeting Laurence Barker at Cranbrook in 1969.[17] In 1973 Tyler accompanied

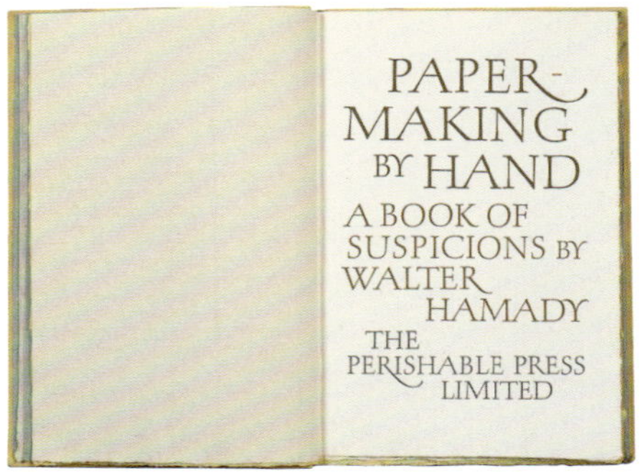

fig. 4 Walter Hamady, Papermaking by Hand, *1982, 11 x 7 1/2 inches*

fig. 5 Winifred Lutz, Epiphenomenon (Phase Transition), *1987, 44 1/4 x 44 1/2 x 105 1/4 inches*

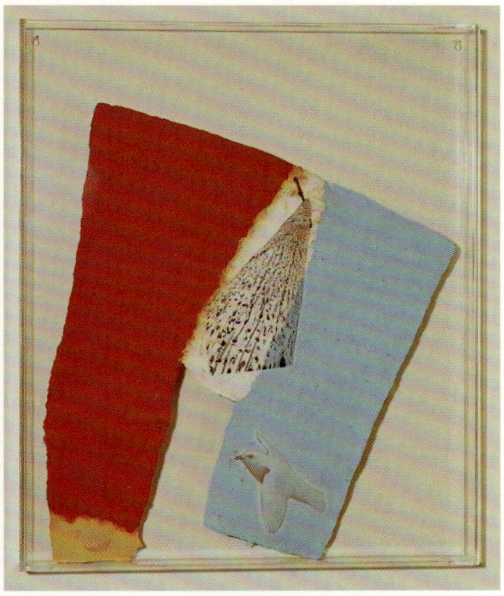

fig. 6 Robert Rauschenberg, Pages and Fuses: Link, *1973-74, 25 x 20 inches*

Robert Rauschenberg to the famous Richard de Bas Papermill in Ambert, France, a mill that had been in existence since the fourteenth century. Their five-day stay resulted in the remarkable *Pages and Fuses (fig. 6)* series in which Rauschenberg explored the process and materials of papermaking and the coloristic possibilities of painting with pulp. Far from being mere exercises, the resulting pieces were exquisite, both visually and conceptually, and Tyler's participation as master printer soon-to-be-master papermaker was clearly critical to the project's success. More than any other project, the acknowledgment of a major well-known American artist such as Robert Rauschenberg making paperworks was highly newsworthy.[18]

After leaving Gemini G.E.L. in Los Angeles in 1974 to move to the East Coast, Ken Tyler continued his paper research and papermaking collaborations with artists and other papermakers.[19] Soon, however, Tyler Graphics Ltd. grew to include extensive papermaking facilities. Tyler's collaborations with artists making paperworks mushroomed and included projects with artists such as Kenneth Noland, David Hockney, Frank Stella, Ellsworth Kelly, Richard Smith, Anthony Caro, Alan Shields, Steven Sorman, and James Rosenquist. True to his nature, Tyler revolutionized possibilities for paperworks by inventing new tools and equipment to work on large-scale pieces and to manipulate paper and pulp in every conceivable way. Whether minimalist, representational, or totally eclectic in subject or style, whether two- or three-dimensional, the challenge was in the process and experimentation. One need only think of David Hockney's *Paper Pools* series of 1978 *(fig. 7)*[20] or James Rosenquist's monumental *Welcome to the Water Planet* of 1988-1989 to appreciate the problem-solving spirit and collaborative excitement enjoyed by Tyler and his artists.[21]

fig. 7 David Hockney working with Kenneth Tyler in 1978 on early paper pulp image, in preparation for the Paper Pools

fig. 8 Louise Nevelson, Nightscape, *1977, 27 x 31 inches*

Other print workshops also incorporated facilities and equipment for paperworks in a variety of ways in the 70s. William Weege, a member of the University of Wisconsin's printmaking faculty, started the Jones Road Print Shop outside of Madison in 1971. Freewheeling experimentation with pulp, papermaking, and printmaking led to an amazing variety of projects with Sam Gilliam, Alan Shields, Sam Richardson, and other artists, with papermaker Joseph Wilfer often joining the collaborations. Weege utilized non-traditional printing techniques in his own work and also incorporated a wide range of paper and pulp techniques.[22] In 1973 Garner Tullis established the International Institute of Experimental Printmaking (now the Garner Tullis Workshop), where he collaborated with West Coast artists including Robert Arneson and Sam Francis on unique and editioned paperworks. Among his important early projects were Louise Nevelson's *Dawnscape and Nightscape (fig. 8)*, black-and-white low-relief paper castings made in 1975.

Atelier Royce, founded by Richard Royce in 1976, was an experimental shop specializing in papermaking and paperworks, though the shop was also set up to work with artists on intaglio and wood-relief printing.[23] Judy Chicago created a large cast butterfly, *Submerged-Emerged*, there in 1976. Claire Falkenstein printed on sheets of black paper and then twisted them into a 1976 sculpture titled *From Point to Cone*. After Ken Tyler's departure from Gemini G.E.L., the Los Angeles printshop continued extensive paper projects with several of its artists, including Keith Sonnier, Frank Stella, and James Rosenquist. A new adventure came for Robert Rauschenberg in 1975, as he traveled to work with papermakers at the textile center of Ahmedebad, India, to create the sculptural *Bones and Unions* series published by Gemini. Although these are but a few examples, printmaking workshops across the country were clearly ready and willing to develop papermaking facilities to accommodate new projects.

As stellar examples of paperworks were created, much attention was given to the surge of new artworks in gallery and museum exhibitions, catalogues, and articles in periodicals and the press. As with printmaking, interest in the medium resulted in a vogue for papermaking that witnessed an unfortunate flood of mediocre works focused more on technical accomplishment than thoughtful process and powerful statement. A negative reaction soon surfaced that threatened to send papermaking and paperworks back to the ranks of craft; however, the masterworks have clearly stood the tests of time. The rest, for good or bad, found their way to the walls of numerous hotel chains, doctors' waiting rooms, and offices where the lure and seduction of beautiful papers continue to find appreciative spirits.

As paperworks and new forms of paper became part of the mainstream of contemporary art, the medium was thoroughly incorporated into the dizzying progression of movements and trends that marked much of the 1970s. Sculptural works by Suzanne Anker, Lynda Benglis, Mary Frank, and George Segal could not be more different, but each artist worked with pulp to create works totally in keeping with their work in other media. Some artists such as Caroline Greenwald remained rooted geographically but incorporated global paper traditions in large-scale paperworks of remarkable diversity. Other artists, as we have seen with Rauschenberg, travel the world over looking not only for papers but the cultures and papermaking traditions of diverse peoples.

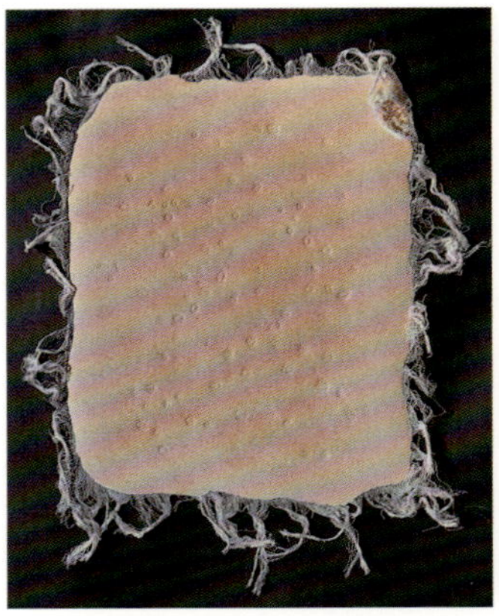

fig. 9 Michelle Stuart, Tsikuping, *from* Tsikomo Suite, *1974-75, 19 x 15 inches*

Michelle Stuart's handmade paper books and prints of the mid 1970s were literally earth art, as she often incorporated earth from the geographical sites she visited into the papers. Her remarkable *Tsikomo* series *(fig. 9)*, created at Tamarind shortly after the famed press's move from Los Angeles to Albuquerque in 1974, were printed paperworks with names taken from the sacred sites of Native Americans of the region. For the series, paper was printed repeatedly to achieve the desired earthen colors and also run through the press with a plate that was built up to create deeply embossed marks. One of the series was then torn and adhered to cheesecloth with wildly frayed edges, creating a textured symbolic object that was quite different from the more straightforward lithographic projects being produced at Tamarind during the 1970s. The growth of the conceptual art movement, which often examined relationships between the visual and the verbal, led to a wealth of books, paperworks, installations, and performance pieces. Paper and pulp were used not merely for their aesthetic and physical qualities, but for the powerfully subversive, humorous, political, and emotional associations that paper and the printed word embody. Before immigrating to the United States, Russian artists Vitaly Komar and Alexander Melamid staged a 1975 performance entitled *The Essence of Truth (Grinding Pravda)*. Photographs of the mute performance show the artists intently cutting the infamous Soviet newspaper into pieces and feeding them through a grinder. The resulting lump of pulp, though still exhibited in museums, holds little interest as an object, yet the performance provided a powerful format for expressing the artists' radical ideas.

Papermills and print workshops with papermaking facilities continued to open as a new generation of papermakers who had trained at earlier shops set out on their own. Artist

Helen Frederick founded Pyramid Atlantic, a center for papermaking, printmaking, and book arts, in 1981. The shop has collaborated with artists nationally and internationally while at the same time participating actively in extensive educational activities in the greater Washington area. The Oakdale Paper Research and Production Facility, developed by Timothy Barrett in the 1980s at the famed Center for the Book at the University of Iowa, has played a vital role in the research and creation of paper and has trained legions of paper specialists. By the early 1980s, Joseph Wilfer moved to New York, and at the request of Dick Solomon of Pace Editions, he soon established the Spring Street Workshop while continuing to work closely with his long-time colleagues at Dieu Donné Papermill. In New York, Wilfer collaborated with many of the Pace gallery artists on paper projects. His collaborations with Chuck Close *(fig. 10)* throughout the decade, resulting in lush pulp portraits in the artist's photorealistic style, are considered among the tour-de-force works in the medium.

fig. 10 Chuck Close, Georgia, *1984, 56 x 44 inches*

New territory was mined throughout the 80s and 90s. As the art world became increasingly eclectic, leading shops continued to find artists with new voices and visions. Of course, as this volume shows, artists who have worked with paper for decades continue to find new ways to explore their ideas. New shops, including the Rutgers Center for Innovative Print and Paper, opened to continue the tradition of experimental work in print and paper within a university

fig. 11 Willie Birch, Million Man March, *1995, 60 x 26 inches*

setting.[24] Artists invited to collaborate at the workshop arrive from around the globe, and the dynamic prints and paperworks works created reflect wide-ranging themes and cultures. In 1995-96 the press invited James Lavadour, Jaune Quick-To-See Smith, and Truman Lowe to create prints and paperworks at the workshop; the diverse results reflect the singular voices of three Native American artists from distant regions of the United States. Artists Ming Fay and Willie Birch *(fig. 11)* have both made large-scale figural images working in collaboration with Gail Deery, each using a variety of papermaking and pulp painting techniques to create dramatically different artworks.

The new narrative movement of the early 1980s lent itself to lushly painted pulp images, and became one of the last distinct "movements" of the century. By the 1990s, the artistic playing field had become totally fragmented – as with everything else in society, the art world was changing rapidly. A mark of an increasingly complex art world was the fact that while many artworks could be understood and appreciated immediately, other works called for interpretation to help viewers understand the concept or circumstances behind the final work. Sarah Brayer's bold pulp paintings of landscape and natural phenomena, such as *Surge* of 1995, seem strikingly abstract and appeal at once to the senses. In contrast, the beautiful, ghostly sheets of watermarked paper made by Lonnie Graham, which hung from the ceiling in a 1995 installation recreating a room in his late father's apartment, are far less comprehensible. The large paper sheets, created at Pyramid Atlantic, were purposely designed to deteriorate and to function as a metaphor for a father who was dissolving before his son's eyes during his struggle with failing memory. Pyramid Atlantic's book projects have also run the gamut between immediately understandable subjects to far more subtle ideas beautifully embodied in book format. Prague artist Jan Cincera's *Homage to the Corrugated Container Board* of 1994 was inspired to transform the most common of materials familiar to all into an elaborate object that begs the viewer to touch the soft colored papers and slowly unfold the accordion structure. The handmade elements of the book's form and manufacture have become a metaphor for the artist's fascination with industrial design and materials.

The history of Dieu Donné Papermill is well documented within these pages, and their scope of projects is testament to the remarkable spirit that has sustained the shop through aesthetic and economic swings. Through twenty-five years of extraordinary production, Dieu Donné has continued the traditions of Dard Hunter, William Stanley Hayter, Douglass Morse Howell, Laurence Barker, and others. Founder Sue Gosin began extensive research and experimentation into the development of rag papers in the 1970s, leading to the creation of unique linen papers with remarkable translucency and similarities to parchment and vellum. Collaborations with artists led to further exploration of non-plant fiber papers. In 1981, the Mill produced 100% silk rag paper for John Cage, paper that was later used by Chuck Close for a series of "spit-bite" prints. More recently, the shop worked with artist Susanna Starr to make pulp and paper out of sea sponge. Gosin's early focus on rag papers placed Dieu Donné Papermill at the forefront of the development of Western-style conservation papers for librarians, archivists, curators and, of course, book and paper conservators.

Dieu Donné's contribution to contemporary book arts is equally impressive. Book projects have often led to intense and long-term relationships with artists and fine art book publishers. Osiris and Vincent FitzGerald & Co. have each worked with the Mill on numerous volumes over the years. Editioned paperworks and other special collaborations reflect Dieu Donné's amazing versatility. Many are aware of a few of Dieu Donné's projects, for example, the ongoing collaborations with artist Chuck Close, which began with Close's long-time friend Joe Wilfer and continue to this day at the Mill with Wilfer's close colleague Ruth Lingen.[25] Other collaborations with artists including Mel Bochner, Sandy Skoglund, and Melvin Edwards have broken new ground and explored new techniques to achieve unique effects.

The Dieu Donné team has embodied the blend of craftsmanship, creative collaboration, scholarship, advocacy, and passion that resulted in major works of art. In addition to collaborating with more than 500 artists and creating thousands of sheets of fine papers, the shop has offered broad educational programs, training, exhibitions, publications, and advice and good will to all those interested in paper. Future projects at the Mill will doubtless continue to break new ground and actively contribute to a continually expanding field of paper art. What does the future hold? Interactive book projects complete with microchips embedded in fiber? Custom papers suited to new printing techniques? Papers and creative applications for new Information Age technologies and equipment? New materials to explore? The possibilities, as always, are endless.

NOTES

1. *While the scope of this overview of twentieth-century papermaking and paper art is concerned with developments in the United States, artists and papermakers throughout the world have been part of a larger international movement. The definitive history of twentieth-century paperworks has yet to be written; however, Sylvie Turner's recent volume,* The Book of Fine Paper *(New York: Thames and Hudson, 1998), surveys the range of contemporary papermills worldwide and describes the papers they produce. Numerous exhibitions and accompanying catalogues surveying the phenomenal paper movement of the 1970s preserve the work of key artists. Notable among these publications are Jane M. Farmer's* New American Paperworks *(San Francisco: World Print Council, 1982) and Jane Glaubinger's* Paper Now: Bent, Folded and Manipulated *(Cleveland: Cleveland Museum of Art, 1986).*

2. *One of the best publications regarding the Arts and Crafts movement in America remains Robert Judson Clark's* The Arts and Crafts Movement in America 1876-1916 *(Princeton, NJ: Princeton University Press, 1972), the catalogue that accompanied Princeton University Art Museum's groundbreaking exhibition of the movement in the United States.*

3. *See Cathleen A. Baker,* By His Own Labor *(New Castle, DE: Oak Knoll Press, 2000), and Dard Hunter,* My Life with Paper *(New York: Knopf, 1958).*

4. *Dard Hunter,* Papermaking: The History and Technique of an Ancient Craft *(published 1943; reprinted New York: Dover Publications, 1978).*

5. *In 1954, the museum moved to the Institute of Paper Chemistry in Appleton, Wisconsin. The Dard Hunter Paper Collection is currently housed at the Robert C. Williams American Museum of Papermaking, part of the Institute of Paper Science and Technology in Atlanta, Georgia.*

6. *David Acton,* A Spectrum of Innovation: Color in American Printmaking 1890-1960 *(Worcester, MA: Worcester Art Museum; New York: Norton, 1990), 92.*

7. *As quoted in* Edna Boies Hopkins: Color Woodcuts, 1900-1923 *(New York: Mary Ryan Gallery, 1986), 2.*

8. *See Joann Moser,* Atelier 17 *(Madison, WI: Elvehjem Art Center, University of Wisconsin, 1977).*

9. *Moser, 9.*

10. *See Alexandra Soteriou,* Douglass Morse Howell Retrospective *(New York: American Craft Museum, 1982).*

11. *See Jan Van der Wateren,* Experimenting with the Book, The Janus Press *(London: Victoria and Albert Museum, 1994). Claire Van Vliet has exhibited widely and received numerous awards throughout her career, including receipt of the prestigious John D. and Catherine T. MacArthur Foundation Award in 1989.*

12. *See Trudy Hansen, Barry Walker, David Mickenberg, and Joann Moser,* Printmaking in America: Collaborative Prints and Presses 1960-1990 *(New York: Abrams; Evanston, IL: Mary and Leigh Block Gallery, Northwestern University, 1995).*

13. *In 1959, Howell custom made large 52 x 72-inch sheets of paper adhered to canvas for Wayne using a process Howell called marouflaying. Wayne painted on these stretched canvases using oil paints.*

14. *As quoted in Jane Glaubinger,* Paper Now: Bent, Molded, and Manipulated *(Cleveland: Cleveland Museum of Art; Bloomington, IN: Indiana University Press, 1986), 4.*

15. *For a complete listing of Howell's students, see Soteriou (unpaginated).*

16. *See Cynthia Clark,* "An Industry Survey: Papermaking Revisited," The Print Collector's Newsletter 10 *(July/August 1979): 80-83.*

17. *The history of Ken Tyler's involvement with paperworks is well documented in Ruth Fine's chapter* "Paperworks at Tyler Graphics" *in* Tyler Graphics: The Extended Image *(Minneapolis: Walker Art Center, 1987), 203-240.*

18. *A 1973* Newsweek *article included a photograph of Robert Rauschenberg working at the Richard de Bas mill and described the variety of the art scene in America. (*"Art Without Limits," Newsweek, *24 December 1973, 73).*

19. *For Ellsworth Kelly's* Colored Paper Images *of 1976, for example, Tyler worked with John and Kathy Koller of HMP Papers. The Kollers had also worked on projects for artists Ronald David and Frank Stella with Tyler.*

20. *See Jan Butterfield,* "David Hockney: Blue Hedonistic Pools," Print Collector's Newsletter 10 *(July/August 1979): 73-76.*

21. *Since "retiring" in 2000, Ken Tyler has continued to work on special projects with artists and is currently establishing a new workshop in Singapore. An excellent survey of Tyler's collaborations on paperworks can be found in* Tyler Graphics: The Extended Image *(Minneapolis: Walker Art Center; New York: Abbeville, 1987).*

22. *See* Jones Road Print Shop and Stable 1971-1981: A Catalogue Raisonné *(Madison, WI: Madison Art Center, 1983).*

23. *Richard Royce has retired and Atelier Royce has been discontinued.*

24. *RCIPP was founded as a printmaking workshop by artist and faculty member Judith Brodsky in 1986, and a papermaking operation and program were added in 1990 under the direction of papermaker Gail Deery.*

25. *After Joseph Wilfer's death in 1995, Ruth Lingen continued the Spring Street Workshop for Pace Prints, though the workshop's name was changed to Pace Editions Ink.*

ILLUSTRATIONS

Dard Hunter
Frontispiece for White Hyacinths *by Elbert Hubbard, 1907 (fig. 1)*
letterpress on paper
7 1/2 x 10 inches (open)

Douglass Morse Howell
America, *early 1960s (fig. 2)*
stenciled linen pulp
17 x 22 inches

Laurence Barker
Untitled, *1979 (fig. 3)*
collage on shaped and perforated cotton rag paper, graphite, ink, printing
17 x 25 3/4 inches

Walter Hamady
Papermaking by Hand, A Book of Suspicions, *1982 (fig. 4)*
letterpress on handmade paper
11 x 7 1/2 inches
published by The Perishable Press, Ltd
edition of 200

Winifred Lutz
Epiphenomenon (Phase Transition), *1987 (fig. 5)*
pawlownia log section, laminate-cast hemp/flax paper
44 1/4 x 44 1/2 x 105 1/4 inches

Robert Rauschenberg
Pages and Fuses: Link, *1973-74 (fig. 6)*
handmade paper, screened images
25 x 20 inches
published by Gemini G.E.L.
© Robert Rauschenberg and Gemini G.E.L./Licensed by VAGA, New York, NY

David Hockney
working with Kenneth Tyler on early paper pulp image, in preparation for the Paper Pools *(fig. 7)*
TGL paper mill, Bedford, NY, 1978

Louise Nevelson
Nightscape, *1977 (fig. 8)*
Cast cotton pulp, pigment
27 x 31 inches
published by Pace Editions, Inc.
edition of 75

Michelle Stuart
Tsikuping, *one of five prints from the* Tsikomo Suite, *1974-75 (fig. 9)*
multiplate color lithograph, embossed on Arches paperwith a plate sclupted with sculpt metal by the artist, hand torn, hand fringed with lithographic cheesecloth laminated on the back of the print
published by Tamarind Institute
edition of 24

Chuck Close
Georgia, *1984 (fig. 10)*
stenciled cotton pulp, pigment,
56 x 44 inches
published by Pace Editions, Inc.
edition of 35

Willie Birch
Million Man March, *1995 (fig. 11)*
pigmented cotton paper with pulp-painting collage
60 x 26 inches
published by RCIPP
edition of 10

Dieu Donné Papermill, 25 Years: The Sacred in the Commonplace

SUSAN GOSIN

Looking back on Dieu Donné Papermill's first twenty-five years, the faces of people, their diverse work, and a studio brimming with activity come to mind. This ancient and yet new process of hand papermaking has provided all of us at Dieu Donné the opportunity to work and play and create in so many different ways with as many gifted people. Although the name Dieu Donné comes from part of my father's name, last year when I met Douglass Howell's wife, Alice, her inquiry about the name and its meaning struck me anew. For the partnership between this sublime natural process and the various people we have had the privilege to share it with is for us at Dieu Donné Papermill – God Given – for in the end, our work and art are the enduring marks of creative inspiration.

This inspiration began when it was first passed on to me, Bruce Wineberg, and fellow students through our teacher Walter Hamady at the University of Wisconsin. The realization that we could not only make our own paper, but that this 2000-year old process had untapped potential as an art medium, was profound and served as the impetus to start Dieu Donné in New York City, where it might serve and be supported by an international art community. Artists often have a knack for recycling and finding the inherent value in the discards of human endeavors. As we were renovating an abandoned industrial loft in lower Manhattan, we were simultaneously rediscovering and reconstructing hand papermaking to fit the twentieth and twenty-first centuries. At the dawn of the electronic age, it is perhaps absurd or at least ironic that something that we use everyday and take for granted such as a sheet of paper could warrant such inspiration.

Papermaking was invented long ago to carry and preserve our most valued thoughts and images. Today, it still serves that purpose, however as the disintegration of contemporary printed matter reveals, there is still a real need for fine, archivally sound paper, whether made by hand or machine. As hand papermakers, we have worked with conservators to address these needs. This traditional aspect of papermaking, making by hand the carrier or passive support for printed word and art, has been the cornerstone of our activities at Dieu Donné. However, by engaging the supporting partner, the sheet of paper, into an active role by custom-designing paper with color, shape, and texture to interact with the surface imagery, we have broken new ground in its potential even as a two-dimensional support. Taking it a step further from its utilitarian history, artists have used the pulp at Dieu Donné with the intention not to make paper, but to make art. With a bow of reverence to our craft and its fine traditions, as the twenty-first century begins, Dieu Donné looks forward to pushing the boundaries of our medium by making art from a process we find both generous and inspiring.

A number of narratives are intertwined in the following pages. There is a story of how making paper by hand was rediscovered in the twentieth century. And how a handful of pioneers each discovered a distinct tributary of papermaking to explore and chart. And for us at Dieu Donné there is the interrelated story of how an inspiration led to the building of an organization and how this organization has reached out to artists and the public at large. Folded into the narrative, there is the story of partnership, which is the very essence of Dieu Donné. Papermaking at its core is a partnership between nature and man. As hand papermakers we have often felt like explorers charting new territory, not with the desire to conquer but willingly led and taught by the subtle mysteries inherent in the process. The other partnership is collaboration. The heart of Dieu Donné is about teamwork and, by extension, our partnership with artists, teachers, students, publishers, curators, conservators, and institutions. Out of this sharing, Dieu Donné has grown from its original role in the American reemergence of hand papermaking in twentieth-century art into a multifaceted nonprofit arts organization.

To trace the roots of Dieu Donné and the revival of hand papermaking, we must first return to the nineteenth century, when papermaking by hand died out in the United States with the advent of the machine age and the use of wood pulp. The rebirth of hand papermaking can be dated to the work and research of Dard Hunter in the early decades of the twentieth century. Hunter's driving interest in fine printing, book publishing, and papermaking led him to assemble and operate the first twentieth-century American hand papermill in Connecticut in the late 1920s. Hunter concentrated his efforts on creating beautiful editions of finely crafted books and prints printed by hand on his paper. Besides his own books, Hunter is recognized for his research and documentation of a disappearing global craft. After Hunter, hand papermaking in the United States relapsed into inactivity until Douglass Howell, home from World War II and motivated by similar concerns, took up where Hunter had left off.

Based on his studies of Hunter's scholarship, Howell built his first hand papermill on Grand Street in New York –

curiously, just blocks from the original location of Dieu Donné. While experimentation and paper production occupied much of his time, increasingly Howell became absorbed with the process as a means of creating his own art. Using historic papermaking techniques that had been invented by diverse cultures for decorative, religious, and commercial purposes, Howell reinvented them as tools of individual expression. With Howell's vision, the papermaking *medium* became the *message* of the artist.

From Hunter's legacy and Howell's inspiration, the seed of hand papermaking took root. Generations of papermaker/artists descended in an unbroken line passing on a growing mastery and knowledge of the field.

As early as 1963, Ken Tyler, working as a fine art printer at the Tamarind Institute in Los Angeles, began a decade of research and development with assorted American and international commercial papermakers to establish standards and create new art papers for a growing fine print market. The development of these new commercial efforts and a renewed interest in paper made by hand intersected when Tyler, at the suggestion of Vera Freeman of Andrews Nelson Whitehead and Elie D'Humeries of Arjormari Paper, rented the historic French handmill of Richard de Bas for a groundbreaking collaboration with Robert Rauschenberg in 1973.

By 1975, there were enough practicing hand papermakers in the United States to convene our first conference, held in Appleton, Wisconsin, at the Paper Chemistry Institute, which housed the Dard Hunter Collection at that time. As about fifty of us gathered to view this historic collection and acknowledge each other's efforts, we sensed the impending movement to come. Of the participants at the conference, many have made in-depth contributions to the field as papermakers, artists, collaborators, teachers or historians, and in business. In their chosen niche of papermaking, each has helped create the structure and substance of our field. Many from this early gathering are now recognized leaders. To name just a few, Kathryn and Howard Clark through their output of fine papers at Twinrocker; Timothy Barrett through his research, production, and teaching at the University of Iowa; Elaine and Donna Koretsky through their scholarship, art, and business; and Garner Tullis through his decades of art collaboration, all contribute to the definition and vitality of the field.

Dieu Donné during the past twenty-five years has touched many of these bases. From the beginning we were motivated to be a part of this hand-papermaking movement because we believed, and still believe, that papermaking offers an almost endless potential for creative expression. Our work over the past two and a half decades reflects this commitment. With an intuited sense of direction, we began at the beginning, mastering our craft, exploring the wide world of papermaking, gleaning what was relevant for this century while marrying contemporary technologies to historic traditions. We did this in the early years by making endless sheets of paper from every fiber or cloth we could get our hands on, making traditional Western sheets, traditional Japanese sheets, then mixing it all up in some strange new hybrid of cross-fertilization that is distinctly American. This balance of mastery and experimentation for its own sake helped us map out the creative territory for our work with artists. While we still actively experiment with new fiber, means of coloring, and papermaking techniques, most of our research today is focused on customized problem solving or project-based activities.

Teaching and training evolved as a natural extension of our ongoing research in sheet production and artistic collaboration. From our early hands-on training of employees and apprentices, a complex program of classes and curricula for adults and children in the varied applications of hand papermaking has developed into a formal educational program. Today, we offer accredited classes to students through the New School for Social Research and specialized workshops for teachers and professionals, and we coordinate classes for children throughout the metropolitan New York area.

As the Mill evolved from a laboratory where we mastered old and new skills of papermaking, we became increasingly involved in artistic applications. This facet of papermaking reached an initial peak of interest and activity in the art community in the late 1970s. Landmark work with established artists such as David Hockney, Sam Francis, James Rosenquist, and Chuck Close was being produced in many papermaking studios throughout lower Manhattan and the country. As the artistic and economic climate changed in the mid 1980s, many of these shops closed or redirected their activities into other media suitable for artistic collaboration. At this juncture, Dieu Donné recommitted its focus to hand papermaking through artistic collaborations, custom sheet production, education, and exhibitions.

The rebirth of Dieu Donné as a not-for-profit in the late 80s brought new support for these ongoing activities. As activities became formalized into programs, the reach of Dieu Donné and hand papermaking extended out to a wider community. An awareness of our role as a vehicle of exchange motivated us to spread the word and show the work. Through lectures, workshops, and conferences we helped link practitioners and interrelated art professionals. Though we had regularly acted as exhibition consultants and curators for off-site exhibitions of work created in handmade paper, the addition of a gallery at 433 Broome Street in 1994 was critical for showcasing this work. Distinct from other galleries and exhibition venues, the focus of our gallery is to show collaborations from our studio and those of our colleagues. As the old adage says, a picture is worth a thousand words – and this has proven to be the case. Our continuing involvement in off-site exhibitions and the shows at Dieu Donné Gallery fulfills an important role in establishing and revealing hand papermaking as an art form. While we have shared many roles with our colleagues – producing sheets of paper, conducting research, and teaching classes – we are neither solely a production mill nor a supply business nor a university facility nor a contract studio nor a publisher. We are unique as a nonprofit arts organization fulfilling our distinct mission in hand papermaking, drawing from and serving people far beyond New York City but very much a part of its vital cultural community.

The following four sections are articles that were originally printed in the 25th-anniversary series of Dieu Donné's PULP quarterly newsletter and are reprinted together here as a summary of Dieu Donné's first twenty-five years.

1974 – 1978

Dieu Donné Papermill, Inc., began as Dieu Donné Press in a 1974 book arts class at The University of Wisconsin in Madison. Bruce Wineberg (my husband at the time) and I wanted to publish an illustrated book of his poems, so I took the fateful walk to the end of the hall on the sixth floor of the Art Building in search of a professor who had the skills and equipment to help me. After a cursory glance at the rows of typecases, printing presses, beater and papermaking equipment, I signed up for Walter Hamady's class. Little did I know how it would change my life; nor did I know that fellow students such as Steve Miller, Ken Botnick, Suzanne Ferris, Neil Bonham – not to mention Paul Wong – would continue to be lifelong friends and colleagues. As an artist and as a teacher, Walter Hamady inspired a virtual army of book art students and future hand papermakers with the magnitude and quality of his work and the sheer vitality of his teaching. It was an extraordinary introduction to the field.

One of the highlights of the Wisconsin days was the first North American Hand Papermakers Conference, organized by Joe Wilfer and held in Appleton, Wisconsin, in the fall of 1975. The fewer than fifty participants from around the country gathered together for a show-and-tell and a chance to see the Dard Hunter Collection at the Paper Chemistry Institute in Appleton. After hearing about the adventures of Howie and Kathy Clark in Indiana, Garner Tullis in California, Tim Barrett in Japan, and the Koretskys in Boston, Bruce and I announced our intention to open a mill in New York in 1976. Richard Minsky, attending the Appleton Conference, told us of his newly founded Center for Book Arts in New York, and we made a date to meet in the following year.

Susan Gosin at 3 Crosby Street

In the fall of 1976, Bruce and I packed our worldly goods and bade farewell to family and friends to begin our new life as Dieu Donné Press and Paper, Inc., at 3 Crosby Street in New York. Before leaving Madison, we spent months researching equipment and materials for our new enterprise, knocking at the door of Wisconsin's commercial papermakers for advice and handouts. We engaged former sculpture teacher Art Schade who, with a troop of friends including painter Jane Hammond, built an oversize laboratory pulp beater to our specifications. As a side note to up-and-coming hand papermakers, who can now call Lee Scott McDonald or Carriage House when they need supplies or equipment, these businesses did not exist yet. The revival of hand papermaking was still in its infancy, so all equipment had to be found secondhand or designed and built from scratch, and supplies were foraged from other industries.

Within weeks of unpacking, we began drawing up plans for the mill on the fifth floor of our loft building in lower Manhattan. The next nine months of gestation were spent consulting with engineers, designing the studio, building

Bruce Wineberg at 3 Crosby Street

equipment, tracking down suppliers, and endlessly testing machinery, rags, fiber, dyes, pigments, and sizing agents. Our first shipments of 100% cotton rag came in 500-pound bales from Gerofsky Brothers. We scoured the garment district learning about different kinds of rag cuttings: shirting, sheeting, even diapers. Dupont and Hercules representatives were both patient and amused by our never-ending questions about coloring, sizing, and wetting agents, and the panoply of chemical additives used in the commercial market to address the problems of colorfastness, lightfastness, durability, and longevity.

To quote installation artist Andrew Ginzel, who helped create Dieu Donné in those early years, "Trial and error seemed to be the order of the day, as Art Schade brought in his battleship of a beater, pigments were tested, fiber sought, and equipment built. The press was built by those accustomed to compressing boxes and garbage rather than museum quality papers, yet it worked. The loft windows filled up with tests for fugitive colors, while the air was strung dense with clotheslines and the latest exploits hanging to dry, trying to overcome the heavy humidity in the air."

Although Soho was still relatively uninhabited, we gathered weekly with artist neighbors and visiting friends for large communal dinners, often followed by a few hours of disco dancing at Studio 54. Every day was an adventure, full of discovery in our field, in the art world and, of course, in the city itself. As Andrew recalls, "Looking back, it was such an extraordinary time when questions begot more questions and the excitement was truly palpable."

By the fall of 1977, we were a fully functioning papermill. With Andrew Ginzel and Ray Holland, we were able to produce hundreds of 100% rag (not linter) cotton and linen sheets each week. Also in the fall of 1977, we participated as hosts and panelists in the second Hand Papermaking Conference organized by The Center for Book Arts for the (now) hundreds of professionals interested in hand papermaking. That number had doubled by the spring of 1978, the time of the third conference, hosted by the World Print Council in San Francisco. Hand papermaking was hot and we were all interested in understanding and defining its contribution to the art world. Besides its creative potential, we sought to define qualities of use and permanence. Participating hand papermakers and commercial papermakers submitted their papers for a full range of tests including strength, durability, archivalness, and application to different mediums. As a result of these tests, all artists and professionals became better informed about issues such as acidity and permanence.

Although from the start we had taken on an unofficial role as educators in our field, we made it official in 1978 by offering the first hand-papermaking class through the New School for Social Research. The Mill was flourishing and our classes were filled, but by midyear Bruce and I had decided to go our separate ways and I continued as sole proprietor of Dieu Donné Press and Paper, Inc. In the fall of 1978, old school friend Paul Wong joined me to continue paper production, art collaboration, and teaching.

1978 – 1989

Two years after relocating to New York in 1976, I found myself owner and operator of a fledgling art business in the guise of a twentieth-century hand papermill. Assisted by a Tiffany Foundation Grant, I was able to ask University of Wisconsin classmate Paul Wong to join me at Dieu Donné. With the turmoil and excitement of the first two years behind me, Paul and I ventured forth to build from our initial research a structure for our interrelated activities in sheet production, two- and three-dimensional artistic collaborations, book publishing, education, and exhibitions. My initiation into the thriving and competitive art world of lower Manhattan was a maturing experience. With Paul's fresh energy, innate wisdom, and unshakable commitment to making art, we broke new ground as papermakers, art collaborators, and artists. Although our business derived from three sources – making custom paper, teaching, and art fabrication – it was the collaboration with other artists that proved to be our driving goal, our continuing source of inspiration, and the fulfillment of our greatest potential.

From 1978 into the 80s, Paul and I worked with many adventurous artists from our own neighborhood of Soho. Over the years, an informal ensemble of artists regularly returned to Dieu Donné to create works that expanded our vocabulary in all facets of papermaking. New York sculptor Barbara Schwartz seamlessly transformed her bold, colorful plaster sculptures into two- and three-dimensional paper pulp multiples. Bart Wasserman, who initiated our artist collaborations in 1977, worked for decades on imagery created by the distinctive interaction of 100% linen and 100% cotton rag pulps. New York painter Virginia Jaramillo explored watermarking as a means of drawing and structuring multiple layers of colored pulp in her evocative paintings. To quote Paul, "Throughout many years of collaborating with Virginia Jaramillo, many Dieu Donné process terms were invented, such as 'double dipping,' 'skinny dipping,' 'throwing veils of

At the vat: Susan Anderson, Gail Deery, Paul Wong, and Virginia Jaramillo

pulp,' 'random couching,' 'pulp painting with the mould,' and 'metal drying' (to simulate a metal surface). Many of these techniques were affectionately identified as the 'VJ Theory,' named after her series called *Theorems*."

We also had the opportunity to work with a wide range of international artists who enriched our world view of the historic traditions of papermaking while bringing their own cultural sources of imagery into play in the making of their art. The Catalan artist, Frederic Amat, who first discovered papermaking with Laurence Barker in Barcelona, brought a distinctly Spanish sense of color, design, and scale to the collaboration, reflecting his strong ties to Tàpies and Miró. In contrast to Frederic's powerful colors and texture, Israeli artist Joyce Schmidt chose to work on an intimate scale using monochromatic colors that reflected her observations of nature and the Negev Desert of Beersheva, Israel.

In 1980 Joe Wilfer, another Wisconsin friend, and one of the initial pioneers to use papermaking as an art medium,

Jaramillo/Dieu Donné Terms

DOUBLE DIPPING:
: dipping the mould into the vat twice

SKINNY DIPPING:
: skimming the surface in a Japanese-style formation

THROWING VEILS OF PULP:
: throwing buckets of diluted pulp onto the mould

RANDOM COUCHING:
: unregistered couching

PULP PAINTING WITH THE MOULD:
: painting with diluted pulp onto the mould and couching onto a base sheet or felt

METAL DRYING:
: pressing a freshly pressed sheet onto a smooth sheet of metal to dry

joined us at Dieu Donné to work with artists and publish new work. Joe enticed numerous artists to "test the water." One in particular, Chuck Close, took to it like the proverbial fish. Joe was convinced that Chuck could realize his portraits, sans paint or printed ink, purely by using paper pulp. After months of fine tuning a grid stenciling system and experimenting with dozens of buckets of colored gray pulps, Chuck and Joe created the first successful paper pulp portrait, editioned for Pace in 1981. Several more increasingly refined images followed at Dieu Donné until the all-absorbing project came under the auspices of Pace Editions. What Joe left behind was his brilliant legacy of ingenuity, determination, and a style for collaboration. Today, Paul Wong continues these collaborations at Dieu Donné with Chuck Close, working with Joe's colleague at Pace, Ruth Lingen, and using new techniques and pulps developed from our twenty-five years of experience.

All along we maintained a symbiotic relationship between craft and art. As Paul puts it, "Through the first artist collaborations, new papermaking techniques were developed at the Mill which in turn stimulated the production of many custom-designed papers we could offer to other clients. Each of these activities worked back and forth to advance and challenge the process and medium of paper."

So Paul and I filled in the gaps between artist collaborations with a kaleidoscope of custom papermaking to accommodate the conservation needs of the Library of Congress,

At the press: Chuck Close and Joe Wilfer

the drawing needs of artists like Cy Twombly, and the printing needs of publishers such as Universal Limited Art Editions. Although most of these fine papers were made from linen and cotton rags, when, for example, Richard Artschwager wanted to simulate commercial Celotex panels, we were delighted with the challenge of using an unusual fiber like sugarcane to satisfy his request.

Many of our most unique sheets of paper were designed to illuminate, as well as support, text and imagery. From its very beginnings in 1974, Dieu Donné Press and Paper has been a vehicle for fine printing and bookmaking in partnership with papermaking. As the business grew, we continued this tradition, publishing new editions while welcoming collaborations with other publishers. Once again the diversity of book projects with artists such as André Kertész and Richard Tuttle reflected papermaking's agility and ability to serve each master uniquely. And in honor of our 10th anniversary in 1986, preservation librarian Mindell Dubansky curated an exhibition of our book projects at the Metropolitan Museum of Art.

Papermaking was the foundation of all of our artmaking activities, but from the first days, our role as educators became increasingly complex. In the beginning, we had to train our staff, and as early as 1977 we started offering apprenticeships. By the early 80s, a handful of other mills had opened studios in downtown New York, concentrating on various facets of hand papermaking such as paper casting and pairing lithography and etching with handmade paper. But even with these other outlets, zealous students lined up for workshops and internships. Although we did not formalize this training into official programs until the late 80s, we taught and advised many of the next generation of papermakers. A few went on to build mills internationally – Octavio Roth in Brazil and Joyce Schmidt in Israel – while closer to home, Andrew Ginzel established a papermill at Cooper Union and Ray Holland started a program at the School of Visual Arts. In particular, the Rutgers Paper Facility, established by Gail Deery, who trained at Dieu Donné from 1984 to 1985, serves as a model for student instruction and professional collaboration today. Our teaching, which began informally as on-the-job training, became institutionalized in 1978 with accredited classes at The New School and has grown into a multilevel network of internships and classes for children, adults, teachers, and art professionals.

As our activities broadened, so did our role in the ever-widening world of hand papermaking. In 1980 Paul and I led a panel on art collaboration at the International Conference of Hand Papermakers in Boston. In 1982 we acted as consultants for and participants in the American Craft Museum's survey exhibition of contemporary handmade paper art. We also served as advisors on various book publications, such as Bernard Toale's *The Art of Papermaking*.

Paper As Paint exhibition at F.I.T.

While Paul's own art as well as our collaborations with other artists were regularly represented in national and international exhibitions, Paul also acted as curator for important shows such as the 1985 *Paper as Paint* show at the Fashion Institute of Technology (F.I.T.) Gallery in New York. These group exhibitions, and Paul's solo shows at Perimeter Gallery in Chicago and Condeso-Lawler Gallery in New York, showcased his extraordinary development as an artist working in the medium. Whether working two-dimensionally in brightly colored pulp or three-dimensionally in large-scale sculptural installations, Paul has continually displayed his unique mastery of the process, incorporating his pioneering techniques with a personal and elegant vision. His experience in collaboration has enriched his own work and those he has touched at the Mill. It is a testament to Paul's sheer creativity, generosity, and equally potent ability to solve problems that such diverse and fine work has flourished at Dieu Donné.

As Dieu Donné stabilized in the mid 1980s, I turned my attention to personal matters, remarrying and starting a family. Paul continued doing his own work and the business of Dieu Donné. In 1987, after numerous discussions about how to structure Dieu Donné's future, we decided that the best way to ensure the continuation of our varied activities in papermaking, collaboration, education, and exhibitions was to change our status to nonprofit. As a new nonprofit organization, serving and receiving support from a wider constituency, once again we looked forward to growth and a greater fulfillment of our potential as papermakers and artists. In 1988, as President of the new Dieu Donné Papermill, I began to assemble a board of directors and hired Jill Bullitt as executive director to help us restructure our organization. With Jill's guidance, we defined our programs to fit our mission and formulated a successful fundraising initiative. Fresh from the completion of a Watson Fellowship in Japanese papermaking, Mina Takahashi came to Dieu Donné in late 1989, first as an intern and soon to be leader of Dieu Donné for the next decade.

1989 - 1999

In the life of every organization there are periods of growth, of stability, and of change. When Mina Takahashi returned fresh from her studies in Japan, the staff and board of Dieu Donné were hard at work establishing an infrastructure and programs to allow for new growth. Mina's timing was perfect. While her Watson Fellowship studies in Japan focused on traditional Japanese hand papermaking, she was also exposed to international artists using the process to make their own art. From research at The Papermaking Museum in Tokyo, Mina compiled a list of contemporary mills to visit on her return to the U.S. Settling on New York, she began her quest, saving what she hoped might be the best for last. As Mina recounted her first impression of Dieu Donné at Crosby Street, I began to laugh, remembering the countless stories told by customers and colleagues of their first attempt to locate us.

Mina Takahashi in front of 3 Crosby Street

As Mina remembers, "It was a crazy rainy day. I finally found 3 Crosby Street hidden away between Soho and Little Italy. It was the last building on the block. Except for the number '3' above the elevator door, there was no signage, just an unmarked intercom. I walked to the nearest phone booth on Canal Street and with great relief connected to an impatient male voice, who nonetheless gave me instructions on how to get into the building. I returned to the anonymous elevator door just in time as it opened directly onto the street. I ran inside as the doors mysteriously closed and it started going up. The elevator stopped but the door didn't open automatically, and there was no one to greet me. After a moment's hesitation, I pushed it open only to find myself staring down a hallway. Turning the corner, I called out and a smiling blonde, Sharon Dürr, peeked her head around a corner wall in greeting. I noticed a lot of shoes lined up where some oak flooring met the studio wet floor. Since I had just returned from Japan where it is customary to remove your shoes, I called out, 'Should I take off my shoes?' and started untying my shoelaces. With water and pulp dripping down their rubber aprons and boots, Paul and Sharon peered around the wall and gave me a quizzical look. I quickly realized then that this particular custom of removing your shoes was to exchange them for protective rubber boots. I immediately retied my shoelaces and introduced myself. In spite of their demanding schedules, both Sharon and Paul took time to hear about my Japanese studies and to carefully look through the portfolio of papers I had made while I was in Japan. Needless to say, I knew this was where I wanted to continue my papermaking, and although there wasn't any work available at that time, I volunteered as an intern that day. In a few months, I became a studio assistant and, as I soon learned, all of us at Dieu Donné wore many hats."

Shortly thereafter, Mina began assisting Jill Bullitt with administrative duties, and together they brainstormed on effective ways to increase our reach and public profile. As part of this strategy, they initiated the newsletter as a means of keeping in touch with our community. They organized a public lecture series, which brought colleagues such as Laurence Barker and Tim Barrett to Dieu Donné to share their work and research. In 1990 they helped initiate an artist residency called the Workspace Program, which provides support for studio collaborations to selected artists. Since 1990 we have offered these funded residencies to forty artists working two- and three-dimensionally in their art.

With critical new initiatives in place, Mina took over the administrative duties of Dieu Donné as program director when Jill left New York in mid 1990. While our ongoing activities and new programs reflected the vitality of Dieu Donné at this time, they also pointed to our greatest problem: we had outgrown our space at 3 Crosby Street. In conversations with Richard Minsky at The Center for Book Arts and Robert Blackburn at The Printmaking Workshop, we realized not only that they had similar space problems, but that together we might be able to find a solution. This began a partnership, eventually called The Fund for Paper, Print and The Book, which took us on a real-estate hunt to house our interrelated organizations under one roof.

In 1991, with our own space at 3 Crosby Street used to full capacity and no new housing in immediate sight, Mina and artist Ken Polinskie devised a mobile educational program called the Responsive Hands Outreach Program to meet the growing needs of children in the New York metropolitan area. Mina and Ken assembled a portable papermaking workshop that fit onto a luggage stroller and traveled by subway on a citywide tour of schools and arts centers. From the first summer camp involving 45 enthusiastic kids, the program has developed into a wide range of classes that integrate science, language, and the visual arts. Today, John Currie, our Outreach Coordinator, organizes workshops for teachers and coordinates instruction for more than 2,500 children each year.

To celebrate our 15th anniversary in 1991, we organized two off-site exhibitions in New York to showcase fifteen years of collaborations. An exhibition curated by Margaret Mathews-Berenson at Lintas Corporation revealed the wide scope of two-dimensional work produced in our studio, and the group show chosen by Mina and Paul for the gallery at The Center for Book Arts concentrated on book collaborations. To complement these exhibitions, Mina coorganized a *Paper Chase* with The Center for Book Arts, Manhattan Graphics Center, and Kate's Paperie, which involved demonstrations and workshops for the public. Also in 1991, the New York Public Library's landmark exhibition *On Paper: The History of an Art*, featured Dieu Donné collaborations and a lecture by Paul on "The New History." Meanwhile, Mina organized a hand-papermaking/paper conservation partnership with the Library of Congress and the University of Iowa to establish definitions and standards for papermakers and conservators. This exchange was a critical step in the professionalizing of this aspect of hand papermaking.

You don't have to read between the lines to realize how quickly we were growing in the early 90s. With Mina's fresh input and accomplished grantwriting skills, we were able to secure an NEA Advancement Grant to develop a long-range plan. Under the guidance of Richard Evans's professional technical assistance, we, staff and board of directors, reviewed our current status, defined our goals, clarified our options, and drew up a detailed plan for growth.

A good measure of luck and timing has often brought just the right person to Dieu Donné, and as Mina recalls, when Helen Hiebert first visited the Mill, initially she came to ask advice about papermaking in Japan. But what impressed Mina was not only Helen's commitment to hand papermaking but her unique vision of how it could be used as an educational tool for children. After taking the New School class, Helen became a part of the Dieu Donné team and within a year developed and implemented her dream of a Papermaker's Garden. Working with elementary school teachers and children in New York, together they planted a flower and vegetable garden, harvested the plants for pulp and made paper for their own books. With our educational programs flourishing along with professional collaborations and exhibitions, in 1992 Helen assumed these responsibilities as program director, allowing Mina the time to manage Dieu Donné as executive director.

Meanwhile in the studio, Paul continued to collaborate with artists such as Leone and Macdonald, helping them pulp and make sheets of paper from the entire contents of their wardrobe for an installation at the Whitney Museum at Philip Morris. Paul also worked directly with publishers Vincent FitzGerald and Ben Shiff to develop unique papers for their book projects with artists as varied as Dorothea Rockburne and Sol LeWitt.

During all of this activity, we continued to look with The Center for Book Arts for suitable housing for our organizations. We received a grant from The Arts Forward Fund,

Helen Hiebert teaching at PS 234

Eugénie Barron with children in the Papermaker's Garden

Ken Polinskie teaching children at 3 Crosby Street

along with invaluable advice and guidance from funders and friends Bob Crane, Terry Savage, and Fredrica Jarcho, to plan our joint venture. However, by early 1994, the urgency of our own space needs took precedence and we began negotiations with artist David Novros for the ground floor at 433 Broome Street. Signing a lease in June, we worked with architect Norman Cox and contractor Curt Royston to design and renovate our new space. Accustomed to making the most of our resources, while 3 Crosby Street was being packed and made ready for the big move, we were still able to house and collaborate with Margita Titlova from Prague in our first international ArtsLink residency.

Meanwhile at our new site, between the comings and goings of miscellaneous carpenters, electricians, and plumbers, artists Marco Breuer and Steven Brower worked on the fabrication of a paper installation for the MTA Arts in Transit Program at the Bowling Green Subway Station. By early fall, colleagues and friends such as Eugénie Barron and Dene Ross filled their cars with packed cartons at Crosby Street and drove the two blocks to Broome Street to unload. At the very end, the hydraulic press was hand rolled over steel pipes along the cobblestone route to Broome Street.

In November 1994, we opened the doors of our new home at 433 Broome Street. And thanks to the extraordinary vision and support of Bob Crane and the Joyce Mertz-Gilmore Foundation, the first and major phase of renovation was completed on time and to our specifications. Collaborations continued seamlessly on a new wet floor four times the size as at Crosby Street. Bart Wasserman, who initiated artist collaborations in 1977, inaugurated the Dieu Donné Gallery with a show of his current work. We opened a retail store, stocking it with professional supplies and handmade paper from our shop and from our national and international colleagues. In celebration, we hosted our first fundraising party, the Grand Opening Paper Mask Ball, with an auction of donated art by more than 100 artists. Working as MCs for the event, artists Gregory Amenoff and Ken Polinskie took turns wooing and working the hundreds of revelers into the night. In spite of our exhaustion from the past months and the excitement of the night, all of us who had become a part of Dieu Donné could feel the good will of our many friends and supporters who had helped us in every way imaginable to realize this shared dream.

With critical support from the New York State Council on the Arts and the New York City Department of Cultural Affairs, work in the studio and exhibitions in the gallery flourished. Beginning in 1994, the six to eight solo and group

433 Broome Street entrance

shows presented in our gallery each year have showcased books, two-dimensional art, sculpture, and installations created by artists working in the studio. Exhibitions of historic and critical import, such as the upcoming Douglass Howell show in June 2001, pay tribute to important contributors in our field. And each year, a fall spot is saved for our annual fundraising event in the gallery, which features dozens of thematic postcards contributed by our growing roster of artist friends. The annual auction and party are often the inspiration for different kinds of collaboration. As in 1995, we invited 15 artists to produce paper clothing for a "Pulp Fashion" show modeled on a runway that ran from the studio to the storefront gallery. A whopping 350 people attended this event.

Grand Opening Paper Mask Ball

Behind the gallery walls, work continued in the new expanded studio in 1995 with artists such as Lesley Dill, on a 90-inch tall *Paper Poem Dress*, and Robert Gober, on giant-size Kleenex tissue papers. Also in 1995, production for a Czeslaw Milosz book began with artists Amanda Guest and Bart Wasserman.

Meanwhile, Mina broadened our reach internationally by attending the International Paper Conference in 1995 in Kyoto, Japan, and by leading tours with Marilyn Sward and Tim Barrett to some of the mills where she had studied. A consultation sponsored by UNIDO to help Indians adapt indigenous plants of the Rajastan Desert in Sanganeer, India, to Japanese style papermaking completed her tour of the East. In another project in 1997, Mina and board member Gail Deery worked with CARE/Ecuador to initiate a papermaking project in the biosensitive cloud-forest region with cabuya farming communities.

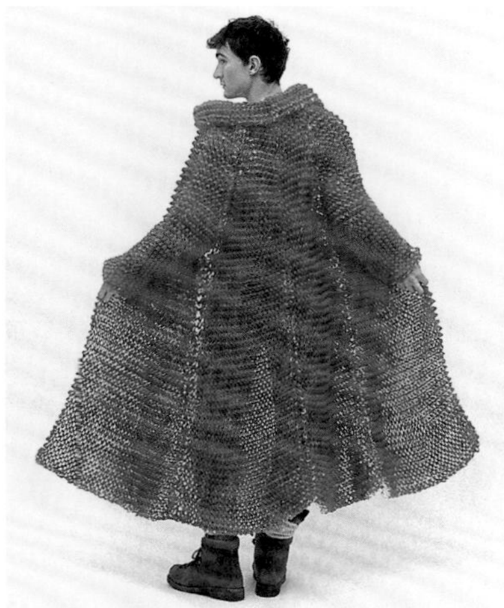

Pulp Fashion: Oliver Herring, Red Coat #2 *(knitted paper)*

To celebrate our 20th anniversary in 1996, we organized two national events. The first, the Symposium of Paper and Paper Art, brought artists, papermakers, publishers, curators, and educators from around the country to exchange views with the public on the state of our field. Following the symposium, we hosted the annual membership meeting of The Friends of Dard Hunter, a national society of hand papermakers. The week's events included a group show of art by Symposium artists held at Cooper Union, eighteen behind-the-scene tours of artists studios, museum archives and publishing facilities, and a treasure map listing 75 paper-related points of interest in the metropolitan New York area. While international consultations and national conferences brought people into direct contact with Dieu Donné, television spots such as Josh Selig's short for Sesame Street increased our public profile.

Mina Takahashi and Gai Deery with Cabuya Paper Workshop participants in Ecuador

With our 20th-anniversary events behind us, we refocused our attention in 1997 on studio collaboration. Under the multitalented supervision of studio manager Pat Almonrode, artists and assistants worked relentlessly to develop and fabricate major paper installations for Sandy Skoglund and Richard Polsky. Pat and his crew cast and printed more than 3,000 paper tiles and a few paper sinks, tubs, and toilets for the Skoglund project *Walking on Eggshells*, which previewed at Smith College. For Richard Polsky's architectural installation at The Jazz Standard, hundreds of paper castings were printed wet, dried, and then assembled at the restaurant. And Mina took a turn in the studio to work with artist Emiko Kasahara to help her draw lines of human hair on editions of handmade index cards and memo pads. Concurrently, Paul collaborated with Arlene Shechet and Donald Baechler on new paperworks. Along with these collaborations at Dieu Donné, Paul was busy assembling a major installation of his own work for the Neuberger Museum in 1997.

Sandy Skoglund and crew on the wet-floor

While all this activity took place on the wet floor, upstairs in administration, Oregon-bound Helen Hiebert seamlessly passed the management of programs to Melissa Potter. As program director, Melissa's responsibilities ran the gamut of coordinating classes, residencies, memberships, exhibitions, and printed materials. Working with John Born and, later, Stacey Lynch, they got the word out through newsletters, announcements, exhibition catalogues, and a catalogue of our goods and services that bore the unique stamp of John Born's humorous writing skills and illustrations. This period of managed growth provided just the right opportunity for Mina and the Board of Directors to begin a three-year study of our activities and options for the coming years. Led again by the indispensable Richard Evans, we reviewed our history, mission, and goals, and with fresh objectivity began to streamline our focus on artmaking.

Pat Almonrode, Ruth Lingen, and Paul Wong in front of "The Crusher"

One event that heralded this renewed focus on art was the generous donation in 1998 by Pace Editions of "The Crusher," a hydraulic press capable of pressing 40 x 60-inch sheets. In the role of curator, Werner Kramarsky partnered with us on a joint project called *Drawn to Scale* to select six artists to create and exhibit new work in the 40 x 60-inch format. These collaborations, later exhibited in 1999 at the Wynn Kramarsky Space and Dieu Donné Gallery, helped create a blueprint for future partnerships with curators, institutions, and the artists we serve. As work progressed on the *Drawn to Scale* collaborations, Winifred Lutz, one of the first and most innovative artists to use papermaking in their work, began an NEA artist residency with Paul. This meeting of peers produced an elegant and provocative body of work for a show that traveled from the Dieu Donné Gallery to sites across the country.

Margaret Mathews-Berenson and Jacqueline Brody at Paper + opening

In the gallery, Jackie Brody selected a group show titled *Paper +* for the Dieu Donné Gallery featuring forty-five artists from Robert Ryman to April Gornik. The range and focus of our guest curator's vision reaffirmed our artistic goals. As the *Paper +* show opened, I joined Jackie and other speakers at the NYU Conference on Contemporary Printmaking. Speaking on the contribution of papermaking in the renaissance of fine prints further clarified our ties as well as our unique distinction in the world of paper art.

Throughout this dense decade of activity and growth, the threads of many individual lives have interwoven into an extraordinary pattern of sustaining relationships and creative work. At the center, Mina and Paul have worked as a team with our dedicated staff and board of directors to nurture an environment that fulfills the potential of our medium as it serves the diverse needs of our community. As leader of overall operations at Dieu Donné, Mina has been the architect of our programs, infusing each activity with a profound respect for papermaking and its potential for artistic expression. Her joy in linking people to a process that helps realize their artistic potential fuels all the human interaction and creative work at Dieu Donné.

1999 – 2001

With each successive collaboration and exhibition, the insights realized during our board retreat in 1998 became concrete reality in 1999 and 2000. Indeed, the perception of hand papermaking as an art form and the identity of Dieu Donné as an art organization were crystallizing and maturing. With fresh entrepreneurial incentive, the board consulted with art development professionals, such as Elizabeth Hahn, to structure and support our artmaking, showing, and collector-building programs.

A new initiative, the subscription publication called the Dieu Donné Editions Club, was inaugurated with a paper/print edition by Robert Blackburn. A copublishing project with Mel Bochner evolved out of the *Drawn to Scale* collaboration with Werner Kramarsky. New work with Donald Baechler, Jim Dine, and Chuck Close was developed with Pace Editions as work for Richard Smith's British Airways Concorde installation at JFK was completed. Alongside these contractual collaborations and Dieu Donné publishing endeavors, Workspace artists completed their residencies, and paper production for artists such as Jamie Wyeth filled out the studio schedule. While the studio was managed by Pat and his crew, Paul was able to take time to concentrate on his own work with the support of a Joan Mitchell Award.

Fresh collaborations from the studio were exhibited regularly in the gallery, as well as a survey show of our fifteen years of collaboration with book publisher Vincent FitzGerald. In honor of this partnership and a gift of his books and prints by Dan and Gayle Weinberg, we hosted an event to present Vincent with the writers, artists, and collaborators in a discussion of their work.

A decade of offering funded residencies to emerging artists through the Workspace Program taught us the value of providing opportunities to artists new to papermaking. Their fresh approach to our familiar process not only produced distinctive results, it enlarged our understanding of its creative possibilities. In 2000 we launched a new residency program called the Dieu Donné Lab Grant Program to introduce a different category of established midcareer artists to papermaking, and projects were initiated with Mel Edwards and Jane Hammond.

We also began three 25th-anniversary commissions with Lynda Benglis, Alan Shields, and Michelle Stuart, artists who had pioneered the use of hand papermaking in their art in the 1960s and 70s. Partnerships with Graphicstudio in Tampa and The Fabric Workshop in Philadelphia brought new work with Kiki Smith and Leonardo Drew, respectively. Artist Christy Rupp worked both two- and three-dimensionally at Dieu Donné on a series of frog sculptures for our 25th-anniversary show and a suite of "labels" for genetically engineered food we commissioned for a group show titled *Environmentally Concerned* at the Nathan Cummings Foundation. At Rutgers University, we coorganized a survey of more than 100 pieces of art, beginning with Douglass Howell and showcasing work of his papermaking descendants, including collaborations from the Dieu Donné and Rutgers archives, as well as an installation of Paul's recent work and a selection of work by current Rutgers students.

To recognize landmark works of art and the artists and collaborators who create them, we awarded Chuck Close and the Pace team of Ruth Lingen, Richard Solomon, and Joe Wilfer the first Dieu Donné Art Award. This year's award will go to publisher Ken Tyler for his outstanding papermaking collaborations with an extraordinary stable of artists including Helen Frankenthaler, David Hockney, James Rosenquist, and Frank Stella.

Keeping in touch with the cultural traditions in hand papermaking, we hosted a workshop and discussion, with the help of Jane Farmer, on Himalayan papermaking with two Tibetan papermakers from the Paper Road/Tibet project.

This brings us to our 25th-anniversary year in 2001. As new work for Alison Knowles and Elena del Rivero is being produced for their respective exhibitions at The Drawing Center and Dieu Donné Gallery, the details of our 25th-anniversary show and catalogue are being completed for its national tour. And while new collaborations with Workspace

Inaugural Dieu Donné Art Award presentation to Ruth Lingen, Richard Solomon, Chuck Close, and Kate Wilfer (for Joe Wilfer), with Susan Gosin

recipients are taking place, work for Lab Grant artists Dorothea Rockburne and Jim Hodges is being developed. Although we continue to produce custom papers, our retail store has gone online to make more space available for studio work. John Currie, with his team of teachers and volunteers, continues to coordinate the off-site educational programs. Meanwhile, Stacey Lynch and Megan Moorhouse manage the myriad details of the newsletter, printed materials, memberships, internships, onsite classes, studio and

Megan Moorhouse and Elena del Rivero on the wet-floor

exhibition schedules, benefit events, and work with the artists. Besides overseeing these interrelated programs, Mina and Paul, along with the staff and board of directors, have participated in a second board retreat. Led by Claudia Chouinard of The Results Group, we digested recent research, redefining our infrastructure to be in alignment with growing activities and our guiding mission. As the new co-chair of Dieu Donné, James Abruzzo is leading us through a strategic planning process to define our program goals for the coming years. And with the guidance of architect and Dieu Donné board member Steve Yablon, we have embarked on a study to plan for our growing space needs.

Twenty-five years ago, hand papermaking was so novel that it inevitably prompted the question, "What is that?" Today, when I teach classes, I begin by asking students, children, and adults, "How many of you have made paper before?" The air is filled with raised hands. Dieu Donné and hand papermaking have come a long way. Our immediate future has a studio, education, and exhibition schedule filled with projects and people whose work exemplifies the very best in our collaborative efforts. With the interest, support, and recognition that has grown exponentially over the years, we look forward to another twenty-five years of partnership with artists, students, teachers, curators, critics, collectors, conservators, publishers, and scholars in hand papermaking.

Therefore, in our twenty-fifth year, we find ourselves, once again poised in preparation for the next phase of growth. As hand papermaking climbs to reach its zenith as an art form, our role as advocates and pioneers of a new medium is being realized as we actively participate in the fulfillment of a dream that began for us a quarter-century ago.

Interview with Paul Wong – 6 December 2000

DONNA STEIN

DONNA STEIN: When did you start working at Dieu Donné Papermill?

PAUL WONG: In the fall of 1978. I came to New York to apply for a job at the Printmaking Workshop in May of that year.

DS Where were you before?

PW I was living in Madison, Wisconsin, where I had met Sue Gosin in graduate school. Sue and her husband Bruce Wineberg moved to New York after we graduated in 1976 to start Dieu Donné Press and Paper on Crosby Street. I stayed in Madison and continued doing my art. About the same time, Bob Blackburn invited me to do a one-week printing project with an artist at the Workshop – it was their way of interviewing. Sue learned she had been awarded a Louis Comfort Tiffany grant to fund an apprentice at the Mill and offered me the job. I had these two appealing choices and decided to work with Sue. Her passion was working with artists, and when I came on board we decided to move in that direction.

DS Was your title Artistic Director or did that come later?

PW For the first year I was hired as an apprentice. The second year, Sue received a grant from the National Endowment for the Arts to pay my salary. About 1983 Sue decided she wanted to pursue some personal goals and start a family, so she left her business under my care and I became the Artistic Director. I ran the Mill with one assistant during the mid 1980s until 1986, when Sue and I began discussing Dieu Donné's direction and how it could survive.

DS Was it initially a for-profit company, or was it always not-for-profit?

PW It was a for-profit corporation until we applied for non-profit status in 1987.

DS How would you describe your role as Artistic Director?

PW In the beginning, we primarily developed custom papers for our clients and maintained an inventory of our paper. We also did a limited amount of collaboration with artists using the papermaking process. Eventually, the activity was divided about 50/50 between producing custom papers and collaborating with artists. We taught a papermaking class through the New School for Social Research as well as occasional specialized workshops. During this period, I was basically running the day-to-day operations. As the years went by, the rent increased, materials got more expensive, and it became harder to meet the financial requirements of running a commercial business, so Sue and I decided to consider alternatives. We discussed various options including partnering with a school or a print publisher. Eventually we decided to apply for nonprofit status to find support for our art collaborations and ongoing research. We assembled a board and hired Jill Bullitt as Executive Director. Because of Jill's expertise in the foundation world, we all learned together how to run a nonprofit organization, so in the beginning it was "learn-as-you-go." The first two years were challenging.

DS And was Sue active?

PW Well, not on a daily basis, but her role as President and Founder was essential to our development. And then Mina Takahashi appeared out of nowhere. She began working at the Mill as an apprentice. When Jill decided to leave, we hired Mina, who had worked with Jill for a while, and with Mina's energy and resourcefulness, we were able to work out the kinks and get the organization on its feet.

DS When was that?

PW Mina started in 1989, so it was around mid 1990. There were a lot of facets of Dieu Donné's activities that we were formulating and initiating. With the help and financial support of the J.M. Kaplan Fund and the New York State Council on the Arts (NYSCA), we started developing programs that directly benefit artists through residencies such as the Workspace Program. Mina and Helen Hiebert, Program Director in the mid 1990s, started developing a wide range of educational programs. We quickly realized that the space at 3 Crosby Street was inadequate; it had many problems apart from being too small. After considering various alternatives with the Board, we found our current space. We had explored the idea of finding a building to accommodate The Printmaking Workshop, The Center for Book Arts, and Dieu Donné, creating a center for paper, print, and the book. After more than a year of research and work with our partners to find a suitable space, Dieu Donné decided to go it alone. We found a new space that would make a dramatic difference for our growth, and with a grant from the Joyce Mertz-Gilmore Foundation we were able to move to our Broome Street address.

DS How has your role as Artistic Director evolved?

PW My role has not really changed; I'm still executing the responsibilities I've always had, but now aspects of paper production and working with artists are divided among two other studio staff members and myself, so everything is not falling on my shoulders, which makes me happy. It allows me to step away a little and gives me time to do my own artwork and deal with my own life.

DS You've outlined the administrative aspects of the organization, but what about your creative role as Artistic Director?

PW Sue and I developed a shared aesthetic philosophy at Dieu Donné for the kinds of paper we make as well as how we collaborate with artists. I think every paper mill or studio has a personal stamp that's identifiable, and I feel my influence is very much a part of how things look. I feel a strong responsibility for maintaining the high standards associated with Dieu Donné. In a conversation the other day with one of our 25th-anniversary commissioned artists, Michelle Stuart, she asked about how we help artists create. We try not to impose our own taste on what's produced here but work within the parameters of each artist's aesthetic. As a result, the art that comes out of Dieu Donné is extremely diverse. Sometimes people can't believe that it all came from the same shop.

DS Earlier you said that you have a certain philosophy and you think that it's manifested itself in a kind of "look" that is associated with Dieu Donné.

PW The art produced at the Mill is extremely varied, but I think there's a certain identifiable aesthetic. Our vision is not craft-oriented but has a high level of refinement. The work is very clean looking and sophisticated. Within that point of view, it is possible to be painterly, minimalist, or conceptual.

DS The word "collaborate" is derived from "co," which means "together," and "laborare," which means "to work." So the definition of collaborate is "to work together with others or another; to cooperate." I associate collaboration with words like trust, exploration, and sharing. How would you define collaboration in terms of the projects you and your staff have produced at the Mill?

PW We've worked with so many artists that the word collaboration has a very broad connotation. Artist collaborations are as varied as the artwork or the personalities involved. Most of the artists who work at the Mill have never worked in this medium or with these materials. Naturally, they rely on me or someone else at Dieu Donné to help them understand the possibilities of the process, the equipment, and the materials to accomplish their goals. For the most part, we're not only the technician who has the skills to create the pulp and form a sheet of paper, but we must understand an artist's aesthetic and personality and be able to problem-solve on our feet as we work. On the other hand, artists who are already involved in papermaking may come to us for assistance in fabricating projects for which they don't have the equipment to produce the particular pulps they want to use, or don't have big enough moulds to make large-scale works. They know exactly what they want to do and guide us through the project. We're almost like extra hands. In some cases we make suggestions in the working process, especially when it involves total exploration of the medium. Everyone on staff is an artist or has done artwork, which I think is a necessary ingredient in performing artist collaborations. It's essential to understand the aesthetics and sensitivity of the artist. It's really challenging, not just as a technician, but also to understand an artist's conceptualization.

DS How do projects actually start?

PW There are many programs at Dieu Donné that define our mission. Over the years the Workspace Program evolved. Each year a panel reviews applications and selects three emerging artists, most who have never worked in papermaking; they might have an interest in paper but don't know what the medium is about. We spend time orienting them, showing them the shop, the materials, examples of fiber and paper, and artwork that has been produced at Dieu Donné and explain what they're looking at. It gives the artist enough information to think about more specific areas of the medium they would like to explore. By extending the residency from five to seven days in the studio, we're now trying to give artists an opportunity to develop a specific project as opposed to just getting their feet wet. We work with more established artists as well, who also know little about papermaking, so it often requires an orientation period. Midcareer artists who come into the studio, like Mel Edwards, often bring years of experience of working in a number of disciplines, so they can apply their knowledge to papermaking.

DS Was Mel Edwards a commissioned artist?

PW No. We have a new program, recently funded by NYSCA and private foundations, called the Lab Grant. It's similar to the Workspace Program, but the Lab Grant is aimed at midcareer artists. Some of these artists might have made prints. Papermaking includes a lot of techniques that have a close affinity to printmaking, especially when you're working two-dimensionally to create an editioned image in paper. There is a long tradition of editioning an image in printmaking. In recent years, the notion of editioning images in papermaking has become established. Variation of an image within an edition is much more accepted in today's art market. Sometimes we work in conjunction with a printer, breaking new ground by combining printmaking and papermaking to create an editioned image.

DS You're not only collaborating with the artists but with publishers and printers. What is your role in the mix between artist, publisher, and studio?

PW For instance, in the recent Chuck Close *Self Portrait Pulp/Pochoir* (2000) (*fig. 41*), I worked closely with Pace Editions' master printer Ruth Lingen to figure out what pulp to use and how to manipulate it with the stencils that Ruth fabricated for this image. Because I have the most experience in working with the materials here, I advise on most of the projects that are going on in the studio.

DS What are the particular challenges of a new project?

PW As I was saying with the Chuck Close project, it's really working with the material and choosing the right fiber, the right beating time, and whatever additives to make it work. There are so many variables that if one aspect of the formula isn't just right, it doesn't work adequately. The staff must be very sensitive and patient, following through with every detail, otherwise the end result isn't satisfactory to the artist. You might have hit on it in a preliminary proof but sometimes it can't be recreated exactly. The collaborator has to diligently follow through, recording what already was done, while maintaining a high level of craftsmanship. Doing something that is totally handmade is very challenging. You can't rely on the machinery to do the work. You can't "slack off." Your mind, your focus, and your skills have to be alert every single time.

DS When you're instructing an artist who is unfamiliar with the medium, how do you react to the way the artist responds?

PW During proofing sessions, the artist and I explore a chosen technique with the appropriate pulps and materials. Depending on the artist's openness to papermaking's natural flow, the latitude of result can be exciting, or unresolved for an artist who is accustomed to other materials and methods. I try to encourage them to stay open-minded during the creative process of working in the studio since there is a transformation from a wet mass into a dried cohesive image. The true test of success is after the drying process, when the artist can evaluate the results. I feel a mutual accomplishment with every artist I work with in sharing some of the discovery and investigation of a process that is not always predictable.

DS Are you able to embrace accidents?

PW All the time. In a way, accidents are kind of built into the whole process. Paper pulp is very fluid, and working in papermaking is not like rolling ink on a plate and getting that kind of accuracy. Occasionally, something happens you don't expect. But it may be to one's advantage, especially when you're working with an artist who is flexible and willing to embrace all the variables, accidents, and possibilities that arise in making something by hand.

DS What is the significance of Douglass Howell's statement "Paper is made in the beater."

PW Obviously, if the pulp you've prepared for a project is not precise there's no use in making paper from it. What Howell means is that to get it right is to get the pulp right. For instance if you need a very short-fibered pulp to make a watermark image, you beat a fine pulp so the formation of the sheet is extremely even and precise without knots that would interfere with the image. Fiber choice, fiber length, and hydration are always critical choices, and it requires years of experience to understand the subtle adjustments for correct pulp preparation.

DS Are digital methods intruding into the technical processes of papermaking? Are they changing the look of a work of art that's made from paper?

PW I don't know if it's really changing the look, but it's helping facilitate certain aspects of artmaking, such as the production of stencils. Utilizing a computerized sign-plotter to cut vinyl film that will be used as a watermark stencil, for instance, has been an incredible tool, rather than trying to cut a design by hand. The use of a photographic stencil also offers a whole new area of image-making in paper. Surprisingly, new technology is often the perfect partner in making a handmade work; it broadens the range of expression.

DS Has there been any effort to computerize the beating or pulpmaking process, so that you'd be able to recreate exactly what you'd done before through machinery rather than through manual operations?

PW We record pulp formulas for specific projects. The Reina beater we have is pretty accurate and can be adjusted to reproduce situations that were done before. Nevertheless, if you don't have exactly the same rag you used previously, it's going to be a little bit different. Duplicating a pulp is not as exacting as getting down to the most finite micron of digital adjusting. It isn't like digital printing where you're dealing with the precise application of ink.

DS Yesterday when I was talking with Alan Shields, one of the 25th-anniversary commissioned artists, he said that though he had worked with handmade paper in different studios, one thing he found so appealing about working at Dieu Donné with Pat Almonrode Studio Director was that he was introduced to many new types of pulp and their potential for use in his work.

PW Because we have the equipment, we look for as many different kinds of fiber as are available. When an artist wants something, if we don't have it we go out and find it. We want to accommodate and stimulate an artist as much as possible.

DS How do you develop special papers to realize an artist's or publisher's concept? For example, Vincent FitzGerald tells you Rumi's story of *On the Art of Painting (fig. 44)* that he was working on with Annette Senneby, and you are asked to develop papers for the cover and the inside of the book. Did you work with the artist, the publisher? How did you invent what was finally realized?

PW In this situation I really worked with Vincent as the publisher and designer of the book. He had already discussed with Senneby her concept for the project and communicated with me what the artist envisioned. He challenges my knowledge and experience to achieve something he can use that may not have been produced before. In *On the Art of Painting*, Senneby wanted the papers to look like rusted metal. I had to create pulp-painted passages with earth- and rust-pigmented pulps dried on metal to simulate a metallic surface. To suggest rust or corroded metal I ended up using a very hard linen pulp to create the veil or curtain that divides the book in the middle. For another project, Vincent asked me to produce a paper that looked wet, so I used pearlescent particles and a Japanese technique of dropping water on an extremely thin layer of pulp to simulate the action of water falling and making marks. In these instances, making a sheet of paper is like creating an artwork.

DS In working on a publisher's project with an artist such as Susan Weil, for example on *Warrior Ant*, where she had an active role in the studio, how does the collaboration develop?

PW When an artist is associated with a publisher and comes into the studio to work on a proof or a B.A.T. (*bon à tirer*), we approach their ideas from a number of different angles or variations to achieve what they have in mind. Artists must wait until they see the work dry to determine which approach is usable or if it is necessary to adjust or modify the process to realize their expectations. It's a step-by-step procedure. Even though there might be a give-and-take in the collaboration, the technician must become "egoless" and completely take on the artist's sensibility and aesthetic. You shouldn't impose your own taste. You have to be sensitive to an aesthetic that may not be entirely consistent with your own. I think that's a really important component in collaborating.

DS As Artistic Director you're let inside the creative spirit of another person. It must be a very intimate experience.

PW Definitely. The artist really does need to feel confident and secure working so closely with someone else. I think a lot of artists don't have that experience. They work in their own studios and for certain people it's a private, esoteric, almost sacred activity. Consequently, it's difficult for some artists to concentrate and create in a public space. Some artists thrive on community and love people around them when they're working, whereas some demand privacy, so the Mill tries to be sensitive to each person's needs and, when necessary, to screen out the rest of the activity that's going on in the studio.

DS If you think back about what you've done over the years and how projects have evolved, are there specific methods that were discovered by Dieu Donné in the papermaking process that have become your "trademark"?

PW We've developed a lot of techniques at Dieu Donné that have influenced others by exposure to the work that's come out of the Mill. One discovery is metal-drying paper, in which pulp is pressed between felts initially and then each sheet is faced with an aluminum plate and stacked and put back into the press again. The paper sticks to the metal as it is dried, creating a mottled, shiny-and-matte surface that suggests metal. A lot of artists have exploited this effect in their work. Other developments have to do with stenciling, such as using a photosilkscreen as a stencil that's laid out on top of the papermaking mould and then dipped into a vat of very thin-density pulp, creating the image in pulp. I know a studio in Europe had developed this approach about the same time we had, although they were not aware of us and I wasn't aware of them until later. It's a staple technique at Dieu Donné. We developed another technique called "blow outs." On top of a newly formed wet sheet we lay a very fine screen that has material attached to it like a stencil – either hot glue, silicone, fabric paint or objects – and use a hose to spray the pulp away from the areas unprotected by the material that's been applied to the screen. You're left with a pulp image after removing the screen. "Blow outs" are an adaptation of Japanese lace-paper, but we exploit it as an image-making technique. When I first started working at the Mill, I developed a way of pulp-painting in my own work using very over-beaten translucent linen pulp. By using nonopacifying pigments, this "pulp-paint" achieves very luminous and translucent color areas that we use currently with artists today. These are just a few techniques we've developed over these twenty-five years that have become signature methods stimulating artists now working in the studio. We have all worked hard to develop a professional facility for the exploration of this unique medium and look forward to offering our skills and expertise to a wider pool of artists as we plan for our growth over the next quarter-century.

Technics and Creativity in Contemporary Papermaking

DONNA STEIN

The founding of Dieu Donné Papermill in 1976 as an experimental laboratory for artists working with handmade paper is a source for celebration twenty-five years later. The Mill's initial purpose of actively promoting hand papermaking as a viable medium through the commissioning and creation of works of art in, on, or with handmade paper has changed little in a quarter-century. By adapting existing techniques from other media and inventing new methods, Dieu Donné has collaborated with artists of every stylistic persuasion to explore uncharted territory according to their own working strategies in the fabrication of paintings, sculptures and installations, unique and editioned paperworks, and artist books. The exhibition and catalogue *Rags to Riches: 25 Years of Paper Art from Dieu Donné Papermill* celebrate and chronicle this abundantly productive history.[1]

Alan Shields

Three Commissions

Three special commissions by Lynda Benglis, Alan Shields, and Michelle Stuart are the conceptual centerpiece of the exhibition and continue a programmatic initiative established when the Mill first opened. These artists were selected both for their history of creating art in hand papermaking and their residence on Long Island, where the legendary Douglass Morse Howell (1906-1994) resurrected the craft and pioneered the first paper art in the United States beginning in the late 1940s.

Alan Shields, who has made innovative two- and three-dimensional handmade paper multiples since 1971 with Joe Wilfer at Upper U.S. Papermill and Ken Tyler of Tyler Graphics among others, has produced an extraordinary group of some forty unique, multidimensional sculptures. In the late 1970s, when Shields worked with master printer and artist Bill Weege at Tandem Press, an adjunct of the University of Wisconsin, he tried dipping wire structures into the paper vat. He found the results interesting, if a little unwieldy, and kept several of these trials in his studio. Shields's commission from Dieu Donné provided the perfect opportunity to depart from the multiple viewpoints of much of his previous paperwork and return to these unresolved experiments. Over the course of two years, the artist has joyously explored relationships of form and color in an inventive and spontaneous notion of three-dimensional drawing.

"I started with wires and screen," Shields recalls, "and began to explore their potential in a number of ways. First, I developed layers of grids to be assembled after dipping and drying."[2] Eventually, Shields used galvanized steel wire and mesh, coated in zinc chromate to protect against rust, to form his armatures, which he dipped into cotton or abaca pulp. Upon immersing the framework into the slurry, pulp collected on the wire, which was pressed or air-dried according to the configuration of the form. Shields was flexible about the results when pulp didn't adhere to the form and was challenged by inevitable imperfections. His ideas expanded in several directions. For him, wall-mounted pieces such as *Bee Bee Wait* (2000) *(fig. 12)* relate to traditional graphics; hanging and freestanding sculptures such as *Mickie's Tools* (2000) invite viewing from all sides; and assembled sculptures like *Stepwell* (2000) *(fig. 16)*, in which the traditional ground is fragmented or missing, have an open-ended quality and change every time you look at them. Stylistically, these multidimensional sculptures recycle ideas that were previously insinuated through layering and weaving paper and canvas, and they successfully reveal spatial relationships, inviting the viewer to enter illusionary space.

As the forms developed, Shields, who doesn't have a favorite color and basically uses hues directly from the tube, experimented with coloration, finally selecting watercolor as his medium. "Watercolor has a direct and intimate interaction with the paper," Shields observes. "It is difficult to paint flawlessly and it's hard to recolor areas, so you have to be decisive. When painting on fresh, unsized cotton pulp, the color seems to be drawn from the brush as if by magnetism. The paper shines through washes of color, resulting in a luminosity that is very special. Watercolor is a fragile medium that doesn't leave a skin to protect the surface, therefore works are vulnerable."[3] The jewel-like colors, chosen in a modified, random method, are distributed in a musical way to enhance the motion Shields is trying to convey. Ideally, the structure disappears and paint is suspended like a "vapor of color in the air." While painting on some of the abaca-dipped forms, the artist realized that in focused light tints would project onto the wall. He hadn't been able to achieve this effect since 1971 when he created *Nina Got it for 100 Francs* on a crocheted bedspread (Collection Solomon R. Guggenheim Museum, New York), where the color seemed to hover in space without any visual means of support. It is only with the abaca-dipped paper pieces such as *The Bullit Train* (2000) *(fig. 13)* and *Sho-Gun-Long-Walk* (2000) that Shields returned to this feeling with success.

Architecture is an important source for this series. In November 1999, Shields visited Barcelona for the first time in twenty-five years, primarily to study the *oeuvre* of Antonio Gaudi. At one of the museums he saw an exhibition of Gaudi's models, some made of string and lead weights, using mirrors to suggest the contours of the Sagrada Familia Cathedral. For *Stepwell*, an inverted ziggurat poised on a plexiglass mirror, not only was Gaudi a reference, but so were stepwells, particular to Gujarat, India, which are constructed so people can walk down into the reservoir to draw water as the water table goes down in the dry months of the year. These stepwells are the social center of small villages because they are cool and have decorative niches for vendors to sell flowers or for people to relax and talk.

Some of the other pieces evolved directly from Shields's three-dimensional canvas forms such as *Ajax* or *Inverted Gumdrop Hanging* (1972-73). *Road Runner Off Ben's Point Replay* (2000) *(figs. 14 & 15)* successfully translates the central circular motif in a 1989-90 watercolor into three dimensions. The pinwheel or firecracker motion insinuates movement and refers to the roadrunner cartoons in which the feet of the cartoon figure spin in a circular pattern. Although this work is meant to hang off the wall, when shown upright there is no denying the allusion to Marcel Duchamp's first readymade, *Bottlerack (Bottle Dryer)* (1914).

Expanding on his interest in the theories of Marshall McLuhan and R. Buckminster Fuller, Shields has investigated the iconographic attributes of plasma physics, which treats the molecular structure of all matter as relationships of abstract pattern. Shields's titles are humorous and clever, consciously trying to deceive. They identify the defining moment and are references for transformation. *Me in 3D* (2000), for example, suggests a self-portrait even though the brightly colored work is a 1-foot hanging cube composed of open squares of mesh in which the rhythmic sequences and color values are related to a grid pattern.[4]

Working at Dieu Donné provided Shields with an opportunity to learn more about different pulps employed in the papermaking process. His pleasure in unorthodox craft and emphasis on hand manipulation have invested each work in the series with its own distinctive personality. These playful constructions deftly realize the implied three-dimensionality of his previous paperworks, once again blurring traditional boundaries between media and challenging preconceptions about color, space, and form.

Michelle Stuart

Although Michelle Stuart has worked with almost every medium and surface, paper has been her primary choice for transmitting aesthetic concerns. Stuart first made paper by hand in the late 1960s when she recycled her own drawings, thereby enriching new work with the memory and material of her earlier efforts. She again made her own paper in 1973 when she decided to include plant life, bones, fossils, and vegetable matter found at particular locales into the actual fabric of her work.[5] No two sheets of handmade paper are alike, and Stuart exploited sensuous and earthy qualities by making found matter intrinsic to the paper and thereby adding another layer of meaning and authenticity to her site-specific artworks.

As part of her residency at Dieu Donné, Stuart created two large artworks on very thin linen paper made to her specifications by Paul Wong. *Zoomorph: Air* (2000) *(fig. 17)* and *Zoomorph: Earth* (2000) *(fig. 18)* continue her studies of natural history and the morphology of change. Inspired by her verdant garden in Amagansett, Long Island, New York, these tableaux are also about place, specifically the creatures Stuart observes that fly and crawl or burrow in her garden. They exuberantly reaffirm her wonder at and love of nature.

Using her standard sequential grid as compositional structure, Stuart assembled twenty drawings configured four high by five wide. She employed a collage concept frequently used in papermaking, embedding drawings between two layers of paper. On the bottom sheet Stuart laid down various thicknesses of string that had been dipped into sepia sculpting wax, freely depicting generic creatures. (You know they are an insect, but not the specific type of anthropod.) On top she placed a second sheet of paper and with a tacking iron modulated the heat, leaving to chance the final disposition of the wax as it spread from the string adhering the two pieces of paper. This working method is similar to techniques Stuart has employed for her seed calendars since 1991 in works like *Aquilegia (Columbine)* (1995-97), a large gridded assemblage of pressed seeds, in which the "DNA juices" of the seeds oozed onto the paper. Also, for more than a decade Stuart has been molding and carving unrefined beeswax as part of the construction of her artworks, so it was natural for her to continue with a familiar material.

Recollected from memory, these gestural drawings are among Stuart's freest expressions, done without any preparation or models. Conceptually, she considers these works a contrast in genders. In *Zoomorph: Air* the imaginary female beings are dimensional, pictured frontally and from the side. The whole tableau literally takes flight. Snails and worms, which have a rejuvenating ecological function in the maintenance of soil, are portrayed in *Zoomorph: Earth* and personify the earthbound male. Within a limited palette and constrained format, Stuart conveys the variety of life, the miracle of individuality. These paper tableaux have led to a new series of embedded paintings in which the abstract linear forms drawn with string are visible between layers of encaustic.

Lynda Benglis

Lynda Benglis, who began making cast paper works in 1978 with Lynn Forgach of Exeter Press and Paper, has created a series of works on two themes. One series of snakelike tubes, which is called *Game Being*, is attached to the wall and appears to be drawings in space. Their titles, *Brother Animals (fig. 20)*, *Game Animal*, *Prey Being (fig. 19)*, and by association their meaning are taken from a suite of prints Benglis made at the PRINT Press, the fine arts press at the University of North Texas, in 1995, using ancient Zuni Gods of the Chase. Some, like *Sacred Dance* (2001) *(fig. 21)*, aggressively extend along the ground, animating the viewer's space. "These are spirits that hold, or not, a sacred or secret space," Benglis explains. "I chose to relate them to the wall and to the floor as if they are living, or had lived, and are but skeletal images of their past and unknown existence."[6] The tubular columns, flaccid and phallic, are constructed of coiled aluminum and brass wiring. Each coil is enclosed by a continuous paper skin of pigmented abaca. After the paper is dry, the coils are crushed and hand manipulated by the artist. The second group of three round, moundlike forms, which the artist calls "Hot-Spots," a term signifying the intensity of form and surface that defines a sculpture, Benglis has titled *Sure Prey I, II (fig. 22), III*. Their title, from Zuni hunting fetishes, suggests the continuousness of life and death. Laminate casting is used to fabricate the moundlike forms. Fresh sheets of translucent pigmented abaca paper are molded on top of a positive urethane foam mould (previously used by the artist in metal casting) and allowed to dry and shrink before they are removed, creating an iconographic impression of mass and surface. Gold-leafed discs beneath these "Hot-Spots," create the illusion of a shadow. For Benglis, the mirage is always gold in the desert.

Always receptive to new materials and processes, Benglis quickly identifies their sculptural possibilities. Her innovative and expressionistic working methods are spontaneous and subjective. Intuitively, she questions accepted artistic conventions and employs chance accidents to her creative advantage. Always thinking on her feet, Benglis finds many ways to achieve the same result. Her works, hybrids of painting and sculpture, weave back and forth between related formal concerns and different media. Like rubber and wax, materials the artist has used in the past, paper is an adaptable chameleon. Malleable and tactile, it can be molded and pigmented, and it has a memory, all factors that appeal to Benglis.

While there are definite sources for the imagery, Benglis did not abstract from something else. Recent bronze sculptures from 1999 such as *Epikos*, *Helios*, and *Medusa*, which like much of Benglis's earlier work fuse form and content, serve as harbingers of the enigmatic mounds. Another visual reference for this installation is a series of glass knots Benglis created in 1993 as an artist-in-residence in Auckland, New Zealand. Molten glass, like paper, transmutes from a liquid state to a solidified structure. These ribbed glassy coils, which suggest the sloughed-off skin of a reptile, celebrate Benglis's direct engagement with elemental forces and natural forms. The color and contour of these ambiguous organic shapes also relate to the organs and systems of the human body. Nevertheless, it is the artist's intention to create a magical illusion of skeletal and crustacean growth and expansion, of exploding and imploding, plasticity and tactility, hardness and softness.

OVERLEAF SPREAD 1, ALAN SHIELDS:

fig. 12 Bee Bee Wait, *2000, 16 x 16 x 5 inches*

fig. 13 The Bullit Train, *2000, 15 1/2 x 15 1/2 x 3 1/2 inches*

fig. 14 Road Runner Off Ben's Point Replay, *2000, (oblique view) 23 1/2 inches diameter x 24 inches*

fig. 15 Road Runner Off Ben's Point Replay, *2000, (frontal view)*

fig. 16 Stepwell, *2000, 9 1/2 x 24 1/2 x 24 1/2 inches*

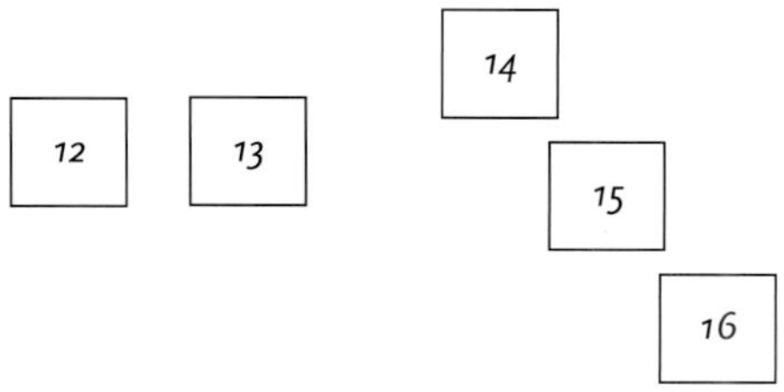

OVERLEAF SPREAD 2, MICHELLE STUART:

fig. 17 Zoomorph: Air, *2000. approx. 14 x 11 inches each of twenty*

fig. 18 Zoomorph: Earth, *2000. approx. 14 x 11 inches each of twenty*

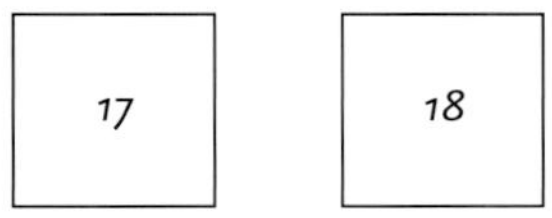

OVERLEAF SPREAD 3, LYNDA BENGLIS:

fig. 19 Prey Being, *2000, 73 x 20 inches x 8 inches diameter*

fig. 20 Brother Animals, *2000, 70 x 15 inches x 4 inches diameter*

fig. 21 Sacred Dance, *2000, 84 inches x 7 inches diameter*

fig. 22 Sure Prey II, *2000, 14 x 20 x 5 inches*

fig. 23 Earth Cavern, *2000, 42 inches x 7 inches diameter*

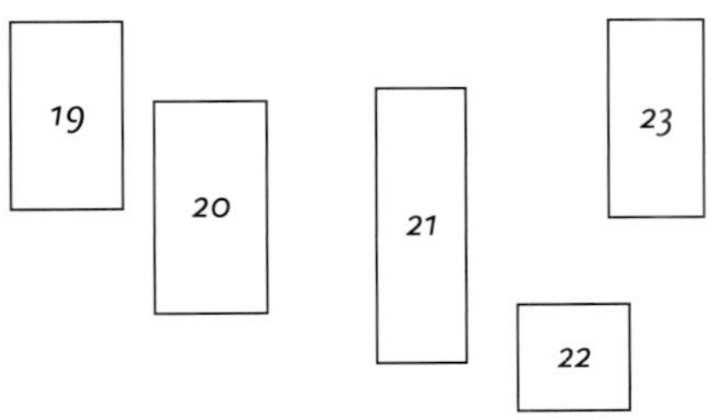

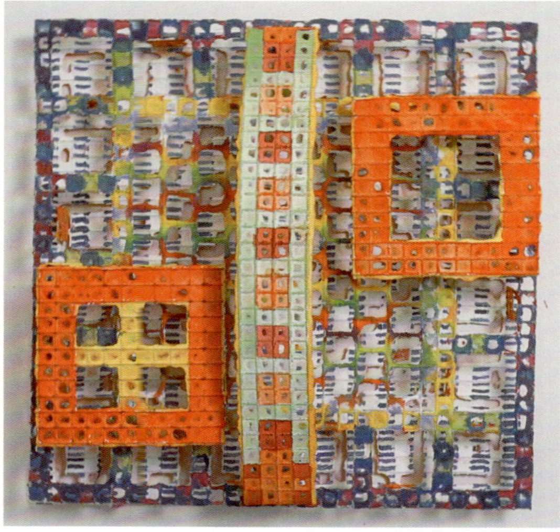

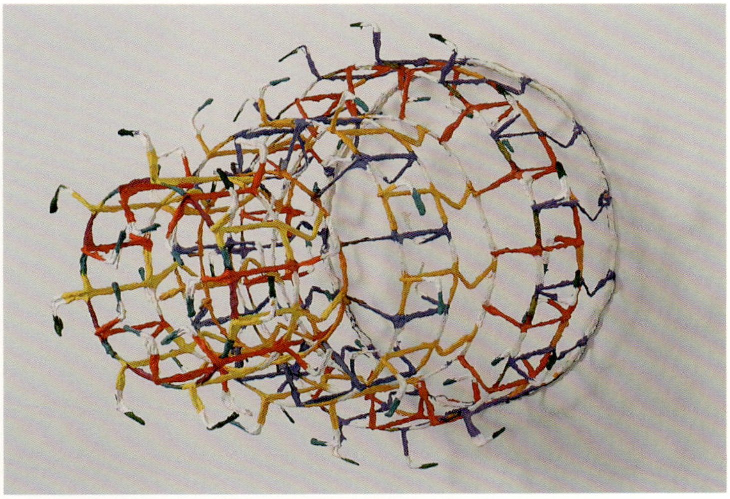

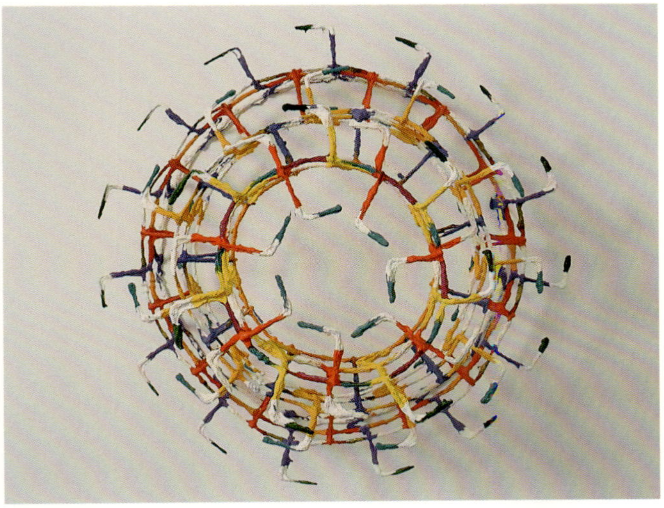

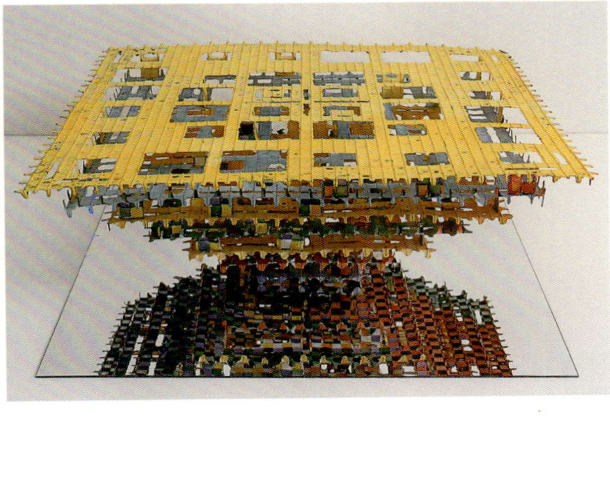

Survey of Paper Art from Dieu Donné

In addition to the three commissions, thirty-six established and emerging artists who have worked at Dieu Donné Papermill since 1976 have been selected to survey papermaking as watermarking, painting, stenciling, collage and embedding, casting, and combined techniques. The selection focuses on artists who have collaborated in the studio to create art in the papermaking process. An intrinsic aspect of papermaking allows the artists to work in both two and three dimensions. These artists' contributions are discussed in different sections of the essay. The chosen works could not have been achieved in any other medium.

Two unusual projects, one by Robert Gober and the other by John Cage, bracket the other artists working in paper, because their work is *about* paper. Paper often imitates something it is not, for example the index cards and notepad of Emiko Kasahara, or the tiles of Sandy Skoglund, both of which will be discussed later in the essay. As part of a recent installation as the official United States artist at the 49th Venice Biennale, Robert Gober produced a suite of three *Untitled* (2000-2001) *(fig. 24)* prints on handmade papers fabricated at Dieu Donné Papermill. They recreate true-to-scale handwritten or typeset texts the artist found, which he carefully recast into etchings. Each of the "clippings" is made to mimic a specific source: a full article from *The New York Times*; a page corner from *The New Yorker* magazine; a xerox copy of a handwritten advertisement. Like his other sculptural objects that are adamantly handmade versions of overlookable, seemingly innocuous things, the handmade papers simulate specific common papers. Different fiber combinations (cotton rag, cotton linter, and abaca) and pigments were tested by Megan Moorhouse at Dieu Donné to achieve a true likeness. Then, Todd Norsten, an artist in Minneapolis, who also is a master printmaker, printed the text from photo-etched plates. Each print was carefully finished to replicate the source. *The New York Times* "clipping" was punched along the bottom and imprinted with a perforated texture on the side; *The New Yorker* was "burnished" by running it between two plates after the ink was dry. This is not the first time Gober has created facsimilies of pages from newspapers and other printed matter. Gober, whose provocative sculptures and installations probe the mundane realities and mythologies of contemporary American life, has contrived more than a sheet of paper. They are reminders

Bomb Suspect's Brother Mutilates Himself

CHARLESTON, S.C., March 10 (Reuters) — The brother of a man charged in the fatal bombing of an Alabama abortion clinic intentionally cut off his left hand with a circular saw to send a message to government agents searching for his brother, the authorities said today.

The Federal Bureau of Investigation said it had received a videotape that the brother, Daniel Rudolph, had made while he was amputating his hand at his home in North Charleston, S.C. The videotape includes an undisclosed message to Federal agents who have mounted a nationwide manhunt for the man charged, Eric Robert Rudolph, 31, the officials said.

Daniel Rudolph amputated the hand on Sunday and drove himself to the nearby Summerville Medical Center. An ambulance crew was sent to retrieve the hand, and Mr. Rudolph was transferred to Roper Hospital, where the hand was surgically reattached, officials said.

Daniel Rudolph has been interviewed by the F.B.I. as part of a manhunt for his brother, but a bureau spokesman said he was not under investigation in the fatal Jan. 29 attack on the New Woman All Women Health Care Clinic in Birmingham, Ala. An off-duty policeman was killed in the explosion and a nurse

The Times Book Review, every Sunday

fig. 24 Robert Gober, Untitled, *2000-2001, 4 2/16 x 5 1/16 inches*

and documents of yesterday's news and propose disturbing connections between art and politics.

Watermarking

Watermarks first appeared in European papers in late thirteenth-century Italy, but it wasn't until 1845 that William Henry Smith invented continuous-tone light-and-shade watermarks, which are commonly used in currency and other security papers.[7] The traditional method of creating a watermark or "cold brand" is to sew or solder a wire emblem onto the papermaking mould, which produces a linear design in the pulp. However, resourceful artists today have experimented with nontraditional approaches, such as cutting an image out of self-adhesive contact paper and adhering it to the mould.[8] In both cases the raised areas create a thinner layer of paper pulp on the mould during the formation of the sheet. Consequently, watermarks are most visible when the sheet is lit from behind or backed with a dark sheet. Numerous artists working at Dieu Donné find the elusiveness of the hidden message within the sheet both subtle and provocative and have used this means to form personal statements within the paper.

Virginia Jaramillo has been working at Dieu Donné since 1979, and the success of her unique approach to handmade paper has been a direct result of her collaboration with and the technical expertise of Paul Wong. *Quadrant-V* (2000) *(fig. 25)* is pivotal in a series of recent paperworks that reestablished a link to driving forces in her work, specifically sacred architecture, archaeology, pre-Hispanic systems of spatial organization, and mathematical concepts such as the ancient "circle of wholeness." She begins by taping a basic geometric design as a linear watermark onto the mould. She uses linen pulp as the painting medium, saturating it with

fig. 25 Virginia Jaramillo, Quadrant-V, *2000, 22 x 30 inches*

fig. 26 Barbara Bloom, Esprit de l'Escalier, *1988, 23 1/4 x 17 3/4 inches*

powdered pigments. A predetermined palette of extremely intense colored pulps is diluted with water and formation aid in separate buckets. In tandem with Wong, Jaramillo throws thin suspensions of colored pulp across the clean surface of the mould in quick succession, each pour affecting the dispersion of the five to six diluted washes. This intuitive and improvisational procedure is repeated several times with subtle color combinations for each layer, until the painting consists of many translucent laminations of varying color layers. The watermarkings used to delineate the drawing affect the separation of areas by displacing pulp during the throws. When the work reaches completion, the sheet is pressed, transferred to a metal surface, and left to dry. The freshness and expressiveness of *Quadrant-V* is the result of spontaneous decision-making by the artist guiding the papermaker's hands.[9] Jaramillo "seeks to translate into artistic terms the structure of our physical, spiritual, and mental worlds from science fiction to science fact."[10]

In 1988 Barbara Bloom, who had never made paper, approached Dieu Donné to produce a suite of seven watermarked papers for one part of an elaborate gallery installation she was planning called *Esprit de l'Escalier (fig. 26)*.[11] Concerned with questions of epistemology and vision, in this case visibility and invisibility and images (photographs) as evidence of existence, Bloom selected pictures of paranormal events such as UFO sightings that she had collected for more than a decade from sensationalist tabloids and proselytizing books. Bloom chose the UFO photos because "if you believe in these phenomena the photos prove their existence. If you don't they are sure proof of fakery. They are perfect indicators of belief."[12]

"The most difficult part of the project," Bloom recalls, "was finding a company who could render a photographic image on a metal plate."[13] Paul Wong led her to the J.J. Plank Corporation in Wisconsin, a manufacturer of watermarks on dandy rolls for commercial papermaking companies, who had previously made the light-and-shade watermarks for Susan Gosin's *The Lectures in Arkady* (1986) *(figs. 27 & 28)* from four-color figurative drawings. They had developed a method of reproducing the light and dark areas of a photograph as levels of high and low relief on a bronze mesh screen by creating a positive and negative bas relief cast of the photographic image; the heated mesh is pushed into relief between the male and female casts, creating mountains and valleys to modulate the pulp. Wong attached Bloom's watermark screens with copper wire onto the papermaking moulds so the cotton pulp would form a continuous tone on the surface of the relief screen. In the center of each sheet, the concealed watermarked images, thinner in the dark areas and thicker in the light, are barely visible to the naked eye. As Gosin has observed, watermarks are fairly easy to edition and make very controlled continuous tone images.[14]

In *The Lectures in Arkady*, Gosin reversed the mesh screens for her light and shade watermarks. This flipping action allows watery pigmented pulp to actually draw the image. For the two-color image of a woman smoking a cigarette and sitting at a desk, Gosin enhanced the white smoke by squirting colored pulp into the wet relief of the freshly couched white watermarked sheet, clearly delineating the design before pressing. For another illustration she applied black pulp to the watermarked area transforming a translucent image of a woman on a bed into an opaque pigmented version. Other watermarks, made from contact paper and representing wolf paw prints in the snow and ripples of water, appear only after repeated viewing and add an element of surprise to the experience of the book.

fig. 27 Susan Gosin, watermark of a woman on bed from The Lectures in Arkady, *1986, watermark: 6 inches diameter*

fig. 28 Susan Gosin, pulped-in image of woman on bed from The Lectures in Arkady, *1986, watermark: 6 inches diameter*

Another approach was devised to make the watermarks in Mel Bochner's *Language Is Not Transparent* (1999) *(fig. 29)* and Christy Rupp's suite *Labels for Genetically-Engineered Food* (2000) *(fig. 30)*. A digital file was sent to a sign maker, who used a computerized sign plotter to cut the image into an adhesive backed vinyl film, which was applied to the paper-making mould. In Bochner's image of thirteen overlapping lines repeating *Language Is Not Transparent*, no line of text stands on its own. His complex layering of the lines of text was achieved in the computer by copying and pasting sections of the line at different densities, which the plotter registered in three layers of vinyl, producing darker and lighter areas of the watermark according to the thickness on the mould. By bonding a black cotton sheet to the back of white translucent abaca, the watermarked words conceptually and physically illustrate their meaning. Rupp's "labels" came out

fig. 29 Mel Bochner, Language Is Not Transparent, *1999, 30 x 40 inches*

of a series of sculptures of engineered insects and vinyl food packaging labels after she learned about the hazards of genetically altered foods. "I was horrified," Rupp admits, "that so much effort has been made by legislators and public relations companies to prevent consumers from learning if products are genetically tampered with."[15] For this project, Rupp decided to enlarge her versions of the same labels in thicker plastic to use as watermarks. She embedded vegetables between two layers of paper the way novel genes are transplanted into fruits and vegetables. Legibility was crucial to her content, and it was a technical stretch to cover thick and chunky vegetables with a watermarked sheet. Rupp combines the imaginary and the scientific. The flattened dried cross-sections of cucumber, onion, and tomato intermeshed with label text appear as a complete packaged commodity.

For the watermark at the top center of each abaca sheet in *Elle Sort Beaucoup* (1999) *(fig. 31)*,[16] Elena del Rivero arranged the letters of her name into a circle, using 3/4-inch-high commercial vinyl lettering adhered backward to the mould. The watermark is a subtle self-portrait reference, placed as it is at the head of the sheet. These "drawings" are a collaboration with Dieu Donné and Werner Kramarsky. Each sheet is identified with a letterpress title. Stains, footprints and torn areas, which are lovingly mended with 1/4-inch strips of linen tape and interlacing stitches, trace their movement

fig. 30 Christy Rupp, Genetically Altered for Your Enjoyment *(onion slices and cumin), 2000, 15 x 22 1/2 inches*

through space and time and show their vulnerability to chance and experience. According to del Rivero, "Paper is a perfect medium in which to recreate the fragility of life."[17]

fig. 31 Elena del Rivero, The Studio *(pearls) from the series* Elle Sort Beaucoup, *1999, 60 x 40 inches*

Painting

Paper pulp is an adaptable medium like oil or acrylic and an intriguing alternative for painting. Pigments and dyes can be added to extend the color range offered by natural fibers. Colored pulps can be poured, sprayed, dipped, and stenciled on the mould during the forming process, by multiple couching and by hand manipulations of the sheet, both on and off the mould.[18] A pulp painting is usually applied to a freshly formed support sheet and fuses into it as one consistent whole. Thus, the support and the medium of the work are one and the same.[19]

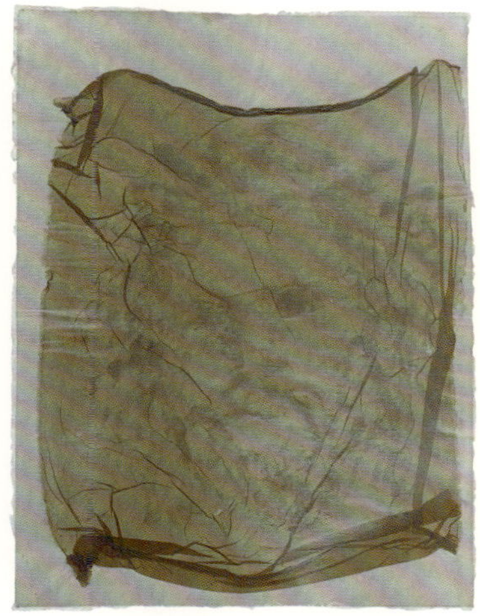

fig. 32 Bart Wasserman, Veronica's Napkin, *1986, 30 1/4 x 22 1/4 inches*

Bart Wasserman was the first artist to collaborate at Dieu Donné Papermill in 1977.[20] From the beginning he has worked only with white cotton and natural linen rag pulp, used together in separate layers but never mixed. His early paperworks were abstract tonal drawings, which he considered studies for his large canvas installations about light and perception. In 1982 Wasserman received an important commission for an exhibition at the Cathedral in Autun, France (near the town of Beaune, where he had lived for seventeen years), commemorating the 2000th anniversary of the city. He began a series of figurative pulp paintings on the theme of the 14 *Stations of the Cross* and over the course of seven years made some 200 images to attain the perfect set. *Veronica's Napkin* (1986) *(fig. 32)*, one of the series, is rendered in a clear, lyrical style in which the religious spirit is palpable. Wasserman used a plastic squeeze bottle to hold the finely beaten cotton rag pulp. He controlled the thickness of the lines by using successively cut plastic bottle tips for squirting. He drew the image directly onto the felt and couched a linen sheet over it. After it was pressed, it was arranged on a freshly couched white cotton sheet. When pressed a second time, folds and creases appeared. When lit from behind, the linen paper acts and looks like cloth.[21] In his collaboration with Amanda Guest for a book of poems, *Czeslaw Milosz* (1996) *(fig. 52)*, Wasserman reconsidered a theme from the Stations and created with a stencil the *Crown of Thorns* on the

fig. 33 Bart Wasserman, Untitled, *1999, 56 1/2 x 75 inches*

title page. In 1989 Wasserman returned to abstract imagery with renewed vigor, increased scale, and expressionistic drama. The pieces are extremely tactile, boldly taking advantage of the cockling or crinoline effect that is created by the tension between linen and cotton pulp drying together. His study of musical composition – counterpoint, harmony, analysis, and structure – has been an important resource for his methods in paper. About two years ago, Wasserman started adding color to the linen paper substrates. For the diptych *Untitled* (1999) *(fig. 33)*, Wasserman used a piece of aluminum shim about 18 inches long as a tool to drag the surface of the cotton pulp to create his image. "I'm not abstracting anything, in fact, I never think of an image when I'm working," Wasserman admits, "I only think about process, amazed by how much of the art is made by the material. I also think in terms of overallness, a concept Jackson Pollock pioneered."[22]

Sarah Brayer's unique paperworks, *Aflame* (1988) and *Up Fifth Avenue* (1998) *(fig. 34)* are part of a series that celebrates New York City at night. *Aflame* was inspired by views of New York Harbor under the bridges in Brooklyn and marks the beginning of the series. *Up Fifth Avenue* is a later interpretation. Both images utilize pigmented linen pulp over a base sheet of natural-colored flax backed with a matching cotton substrate. The double-layered base sheet was Paul Wong's innovation. The tight-fibered surface of the flax lets the pulp drain slowly so the artist has more image-making time. The cotton base produces a smoother finished sheet when dried. "The dark base which I have come to use with my paperworks acts as a tinted ground for the colors poured on top," claims Brayer. "I find this warm, middle value adds a richness to the overall color. I particularly like using the over-beaten linen as my paint as it holds a lot of pigment and yields rich, smooth color."[23] Brayer worked both images very wet. The skies were done first as big splashes of color, which were then manipulated by hand and a fine spray of water to let the pulp bleed and soften. In *Up Fifth Avenue*, the white and orange pulps swirl to create the feeling of movement and light. The buildings were added next. The darks in this paperwork were drawn using a squeeze bottle. The small accents of color in the tail lights and buildings were the final touches. The sheets are pressed twice, the second time with metal next to the paper to give a polished surface to the finished image. Since 1986 Brayer has worked in a paper factory in Fukui,

fig. 34 Sarah Brayer,
Up Fifth Avenue, *1998,*
28 7/8 x 15 1/2 inches

Japan, with a team of professional women papermakers, adapting techniques used in Japanese papermaking. When working more abstractly, she finds a link with Zen ink painting, because the line and form of the gestures are more important than their literal meaning.[24] "Part of what is so wonderful about working in the medium of poured paper," according to Brayer, "is that the pace of the work and rhythm is stream of consciousness. The images are literally pouring out and I'm not always aware of where they are coming from."[25]

One of Eyal Danieli's photographs was the source for *Bent* (1999), the figurative work he made by drawing with pigmented pulp out of a squeeze bottle directly on a wet base sheet. It gives the papermaking medium the same immediacy that drawing with a brush and ink affords. Richard Tuttle combined pulp painting with relief stamping for the illuminations of Mei-Mei Berssenbrugge's poems in *Hiddenness* (1987) *(fig. 35)*. Here, the artist directed the human action needed to create each image and choreographed the repeated motions the papermaker used to edition the work.

fig. 35 Richard Tuttle, Hiddenness *by Mei-Mei Berssenbrugge, 1987, 16 1/4 x 11 x 1 1/4 inches (closed)*

After attending a summer workshop at Dieu Donné Papermill in 1982, Ken Polinskie began collaborating with Paul Wong on a series of paperworks inspired by watercolor and ink drawings. In 1985 Polinskie began a series of self-portraits combined with still-life formats. Humorous, but with a central theme concerning loss and renewal, they were in part a response to the burgeoning AIDS crisis. *Still Life with Self Portrait (Forbidden Fruit)* (c. 1988) *(fig. 36)* is a key example thematically, as well as technically. Polinskie used a direct, fresco-like approach with several unusual aspects. The image and the actual support sheet are created simultaneously with pigmented fibers, resulting in deep color saturation. Working intuitively, Polinskie allows the developing image to determine the proportion and size of the finished work. Large sheets of Styrofoam are used as temporary supports on which the finished pulp painting is air dried, after which it is peeled or "popped" off the foam support.[26]

In collaboration with the Lower East Side Printshop, Dieu Donné Papermill recently published *Cries and Pain* (2000) *(fig. 37)*, a technically elaborate mixed-media print by Juan Sánchez. This formally cohesive and sentimentally expressive composition depicts Sánchez's wife and baby daughter. Like the arched tops of his paintings, the black cotton base sheet was formed and couched to reference the shape of an altar. The artist then threw washes of diluted pigmented linen pulp onto the mould, which was couched onto the black sheet. With squeeze bottles, he freely applied green, blue, yellow and red pulp paints on top of the black sheet. The red strips on both sides of the sheet were formed by pouring onto the sectioned-off mould and couching onto the base sheet and were eventually overprinted for brilliance. Sánchez calls his art "Ricanstructions," a term taken from a 1977 recording by the salsa and jazz percussionist Ray Baretto.[27] The sophisticated layering of techniques and images reflects the artist's awareness of the complex ground from which his imagination springs.[28] Sánchez passionately reconstructs his own history as well as Puerto Rican life and culture in New York.

fig. 36 Ken Polinskie, Still Life with Self Portrait (Forbidden Fruit), *c. 1988, 42 1/2 x 34 inches*

fig. 37 Juan Sánchez, Cries and Pain, *2000, 27 1/2 x 53 inches*

Stenciling

Technically, stenciling and watermarking are two sides of the same coin, because they both affect the density of pulp. In a translucent watermark, the image is invisible, hidden inside the sheet and revealed when held up to the light. With stenciling, colored veils or layers produce the image. Heavy acetate or mylar stencils make it possible to reproduce a complicated pulp image for editioning purposes. Stencils can be marked with permanent ink to register information like orientation and the outer edges of the sheet. For complex designs, it is important to make a clear listing of the order of the stencils, which usually are rinsed after every use. Stenciling, pulp painting, and transfer printing are ways of imposing an image onto a newly formed sheet.

Joe Wilfer joined his old colleagues from the University of Wisconsin at Dieu Donné in 1980 to promote new collaborations with artists, and he managed to persuade a reluctant Chuck Close to create his images in the papermaking process. Wilfer researched various methods of translating Close's portraits into paper pulp images, a daunting task

fig. 38 Chuck Close, Keith I, *1981, 35 x 26 3/4 inches*

fig. 39 Chuck Close, Self Portrait/Pulp-Pochoir, *state proof 2, 2000, 24 3/4 x 19 1/2 inches*

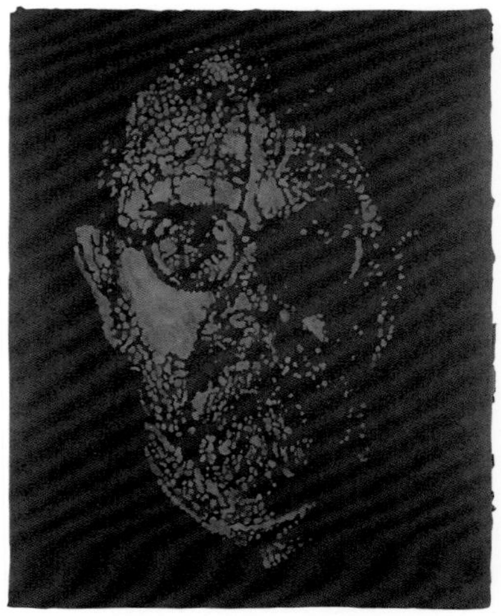

fig. 40 Chuck Close, Self Portrait/Pulp-Pochoir, *state proof 4, 2000, 24 3/4 x 19 1/2 inches*

fig. 41 Chuck Close, Self Portrait/Pulp-Pochoir, *2000, 24 3/4 x 19 1/2 inches*

given that cotton paper pulp lends itself comfortably to abstract imagery, but, at that time, detailed figurative imagery was difficult to render in pulp.[29] Wilfer interpreted Close's systematic vision with a plastic grid template, numbering individual units to receive twenty-two shades from black to white. In what Close rates as "a textbook example of the ideal collaboration between artist and artisan,"[30] *Keith I* (1981) *(fig. 38)*, from a series of five portraits of his friend, was the first in more than a dozen editioned handmade paper images completed during the early 1980s. After a cotton pulp background sheet was formed and adequately drained, the numbered and color-coded template was placed over it into which a gray scale of pigmented pulps from white to black was deposited into designated areas on the grid, adhering to the base sheet. The grill was removed and the completed piece was pressed or, in other editions, left to air-dry naturally. Close actively worked at Dieu Donné, responding to the sheets as they were made, learning and exploiting the technology to understand the process to serve his vision and art.[31] *Self Portrait Pulp/Pochoir* (2000), published by Pace Editions, Inc., and produced at Dieu Donné, was executed entirely in the hand-papermaking process by artistic director Paul Wong, and Ruth Lingen and Mae Shore of Pace Editions Ink. Seven Lexan stencils were used in a method conceived by Ruth Lingen, who adapted enlarged photocopies of state proofs from a reduction woodblock print. On top of a wet sheet of black cotton paper, pigmented linen pulp was spooned onto the exposed areas of each stencil *(figs. 39 & 40)* to create the finished image *(fig. 41)*, which was pressed and dried in an edition of forty.

In 1988 Joe Wilfer, by then affiliated with Pace Editions, also introduced Donald Baechler to the potentials of hand papermaking. For the recent edition *Pompeiian Flowers* (1997), a brass shim stencil originally cut by Joe Wilfer was used. The ambiguous edge handmade paper offers was accentuated by manipulating a shim stencil to achieve an irregular contour. The backgrounds have a mottled effect that varies throughout the edition. "Cookie-cutter" brass shim stencils were also used to create a suite of six handmade-paper multiples called *York House Suite* (2000) *(fig. 42)*, which were based on a group of 126 drawings of the same name made in 1986 at the York House Residencia, a small hotel in Lisbon, Portugal. An artist book was published with a text by Demosthenes Davvetas for an exhibition in 1988 at the Anders Tornberg Gallery, in Lund, Sweden.[32] Ruth Lingen loved the book and wanted to translate some of the images into editions, first trying woodcut and then paper pulp, which Baechler preferred. The delightful and odd little characters in blue and white pulp closely follow the original black India ink drawings on hotel stationery and napkins. Because the original drawings were rinsed with water under the hotel sink faucet, they turned blue. To match this effect three different blue linen pulps were stenciled onto the white cotton pulp carrier sheet.

fig. 42 Donald Baechler, print 4 from York House Suite, *2000,*
12 1/4 x 9 1/4 inches

Barbara Schwartz invented an innovative method for constructing handmade paper images, using multiple layers of stenciled linen pulp on a cotton base sheet. For *Hexagon* (1999) *(fig. 43)*, she cut shapes out of Artcor, a plastic version of Foamcore, and laid them individually onto the papermaking mould. To get a very clean edge, she poured pigmented linen pulp inside the open areas of the stencil, letting it drain, then couching each part of the design onto a cotton base sheet. The different stencil shapes are laid on top of each other and eventually pressed into one cohesive image. In a conversation with Joe Wilfer, Schwartz likened the process to "fresco but happens presto." Wilfer then labeled works produced in this manner "prescotypes."

fig. 43 Barbara Schwartz, Hexagon, *1999,*
20 1/2 x 23 inches

Swedish sculptor Annette Senneby's book project *On the Art of Painting* *(fig. 44)*, by thirteenth-century Sufi philosopher and poet Jalaluddin Mohammad Rumi, is an interpretation of the Platonic dissertation on art and illusion and was a commission from Vincent FitzGerald & Co. The artist conceived a single, physical line mapping the direction for the reader from the front cover to the back, which Wong translated into paper. Using a simple mylar stencil, Wong created a continuous line with white pigmented pulp on the paper liners inside the box and end papers. He also designed a special rust-colored linen paper dried on metal for the center of the book to evoke the curtain separating the Chinese and Greek painters that Rumi describes in his parable.

fig. 44 Annette Senneby, On the Art of Painting, *1989,*
8 x 10 1/4 x 1 1/8 inches (closed)

Arlene Shechet was first invited to work at Dieu Donné in 1995-96. Once she realized paper could be a kind of layering, another skin, she became excited at the possibility of expanding her engagement with process and translating her visual iconography of Buddhist symbols, meditation mandalas, and stupa plans into the new medium.[33] *Mirrored Sites* (2000) and *Hovering Lotus* (1999) *(fig. 45)* are two works from her *Flow*

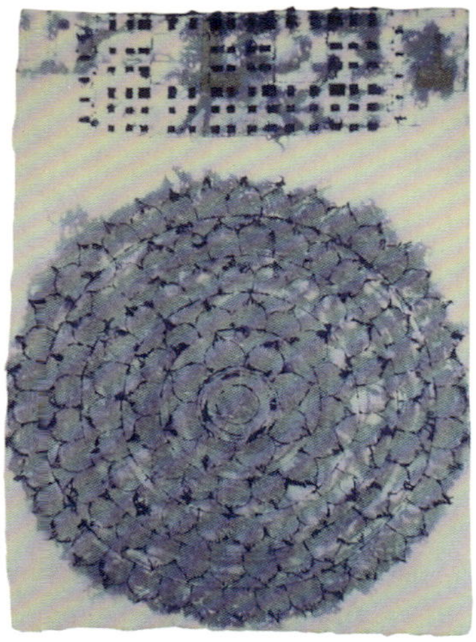

fig. 45 Arlene Shechet, Hovering Lotus, *1999, 35 x 25 inches*

Blue series. Shechet began by drawing the plans of ancient and often destroyed Buddhist temples from sites all over Asia onto mylar, which she cut into stencils and brought to the Mill. On prepared sheets of abaca, she manipulated various blue linen pulps through the mylar, sometimes adding blocks of color or allowing the blue pulps to seep under the edges of the stencil. Because it is important to the artist that evidence of the process of making the paperworks be embedded into the finished sheet, she changed the consistency of the pulp, making it thicker and thinner to imply various levels of dissolution of information. In *Hovering Lotus*, Shechet layers and rotates architectural blueprints, turning concentric bands of repeating arches, columns, and other forms into a two-dimensional mandala. "As 'blueprints,' these images of places," Shechet explains, "combine the architecture of the mind (mandalas) with real site plans of ancient Buddhist shrines (stupas)...In juxtaposing mandalas and stupas, I build new forms using an ancient framework."[34] While the use of blue pigments refers to water, this series is named for a special blue and white porcelain, *Flow Blue*, which is iconic in cultures of the East and West, but originally invented in China. By translating her interests in mathematics, maps, diagrams, and directional elements, Shechet attempts to create a visual language that exists in the West and simultaneously refers to the East.

Mary Judge was invited to work at Dieu Donné as a participant in the Workspace Program for emerging New York State artists. The symmetrical structural format and pouncing technique used in fresco painting, which Judge appropriated using dried pigments to transfer drawings onto her ceramic sculptures, helped the artist determine her direction in papermaking. Judge made mylar stencils, which she placed over her drawings, tracing the lines with silicone. At first, a low relief embossment resulted from the design on the stencil when it was placed face down on a newborn sheet. To achieve more surface detail and interest, she added graphite powder and various metallic pigments onto the stencil surface; some mixed with the water, some resisted, forming air bubbles and other fractal patterns on the cotton pulp. Judge combined four smaller sheets to create her *Heaven and Earth Series #4/101* (2000) *(fig. 46)*. The concentric pattern evolved naturally, transforming the image through repetition.

fig. 46 Mary Judge, Heaven and Earth Series #4/101, *2000, 63 3/4 x 49 3/4 inches*

The paperworks of Margaret Lanzetta, such as *Water Rustle II* (1999) *(fig. 47)*, are a natural extension of her paintings and translate the conceptual and technical processes of layering, accretion, and addition through silkscreen, frottage, and stamp printing into a paper medium. Her geometric patterns are derived from physical materials or photographed sources, which she manipulates digitally to achieve the right scale and proportion before transferring to nylon photo silkscreens and mylar stencils. Pigmented cotton pulps are formed on the screens then couched layer upon layer, wet on

fig. 47 Margaret Lanzetta, Water Rustle II, *1999, 30 x 22 inches*

He used a water hose to wash out extraneous pulp not protected by the objects on the mould. When the objects were removed, a shadow or silhouette of each form remained and these were couched on a linen base sheet. In the preparation of the cotton pulp for *Everyday Connections* (2000) *(fig. 48)*, a

fig. 48 Melvin Edwards, Everyday Connections, *2000, 28 1/2 x 37 1/4 inches*

polyester fiber was accidentally mixed into the brew, which fortuitously created a rich velour-like surface texture. The relief effect is further enhanced by using two fibers (cotton and linen) because of the way they shrink as they dry. When the base sheet for *Le Donné* (2000) *(fig. 49)* was couched, it came off the mould in shreds. Rather than throw it away, Edwards was challenged to make it whole and pulp-painted the bottom of the sheet, considering the opportunity a gift the process offered him. The richly contextual quality of *Everyday Connections*, which includes chains referencing the artist's African-American heritage and the square, ruler, and welding mask the artist uses making sculpture, covertly suggests a self-portrait. Some of the other imagery, such as

fig. 49 Melvin Edwards, Le Donné, *2000, 22 x 16 3/4 inches*

wet, to create the composition. Color functions simultaneously as substance, tonality, mood, and depth.[35] Once the pulp screening is finished and placed on a cotton base sheet, the entire work is pressed to extract water and then dried on aluminum panels to achieve a polished, matte surface. Lanzetta is particularly interested in the continuity and repetition of visual systems that have traveled both in time and place. A good example is an ancient Chinese design of interlocking circles, which reappears in year 2000 in the form of a plastic grid to place over fluorescent light box fixtures sold in Chinatown on Canal Street in New York. Her recent paperworks show intersections of anonymous industrial patterns and nature, such as the form of a snowflake repeated as the imprint on the bottom of a green plastic strawberry basket. Lanzetta makes a metaphysical link between culture and nature in her search for intimacy through seemingly impersonal structures.

Melvin Edwards, a sculptor who has extensive experience with printmaking techniques, inaugurated the new Dieu Donné Lab Grant Program, which introduces midcareer artists to the papermaking medium by offering a twelve-day residency at Dieu Donné. Edwards finds papermaking freer than sculpture and more like painting with washes and solid areas. For the paperworks in Edwards' series *Agua y Acero en Papel* (Water and Steel in Paper), the artist collaborated with Studio Director Pat Almonrode in refining a "blow out" method instead of a stencil, which is conceptually similar to a photogram. Edwards arranged his compositions by placing three-dimensional objects, such as chains, knives, nails, work gloves, locks, and tools, onto a freshly formed but not yet couched cotton rag sheet protected by a fine mesh fabric.

working tools, are straight out of the artist's poignant experience in Angola. Shadows fix the moment, and the deeply human impressions of our existence, such as apparitions of man's inhumanity to man at Hiroshima or nature's power at Vesuvius and Pompeii, inform Edwards's powerful works.

fig. 50 Jim Dine, Red Pants II, *1999, 55 3/4 x 34 3/4 inches*

Jim Dine's large, masterful nine-color print, *Red Pants II* (1999) *(fig. 50)*, was published by Pace Editions, Inc. and Alan Cristea Galleries in an edition of twenty. It combines soft ground etching and drypoint with hand painting by the artist on handmade paper to capture the charm of Pinocchio. To make the paper, Paul Wong designed a mylar stencil like a jigsaw puzzle with taped together sections, which he placed on top of a wet cotton base sheet. By lifting separate areas, Wong was able to apply colored linen pulp and then cover and protect the part he completed until he blended all the edges with a spray mister. The story of the naughty wooden puppet that comes alive for his maker is suggested by the artist's depiction of the eyes, nose, and stance of the figure.

Eyal Danieli employed transfer printing for *The Future is Ours Comrade* (1999) *(fig. 51)*. He drew with water-soluble crayon onto a pellon and laid the image face down onto a freshly pressed sheet of paper, lightly spraying the drawing with water from behind to transfer the image. Each monoprint transfer looks different because the water and pigment movement is unpredictable. The source for this work is a newspaper photograph of an actor depicting a Nazi officer saluting. It is part of a larger series, in which Danieli literally and figuratively blurs the identity of the saluting figure, and refers to Anselm Kiefer's 1969 self-portraits called *Occupations*. The title is a tongue-in-cheek stance regarding the horrors that still await us.

fig. 51 Eyal Danieli, The Future Is Ours Comrade, *1999, 24 x 18 inches*

Collage and Embedding

Collage is the technique by which fragments of pasted materials are assembled to make or enhance a composition and break out of the two-dimensional boundaries of the picture plane. It is a means of incorporating reality without imitating it. In papermaking, an unlimited number of items can be added to the pulp before forming or embedded between layers during couching, including other pieces of freshly formed paper, mica, threads, hair, twigs, or feathers, etc. Collage is a way of bringing art and life closer together.

Amanda Guest's internship at Dieu Donné in 1994 initiated her subsequent involvement with handmade paper. She had been using a single Star of David image in her work when she found a grid diagram in an Islamic pattern book showing a repeated design of the six-sided star. Guest started making embedded nail drawings in handmade paper and watermarked images with the grid pattern. In 1995, founder Susan Gosin saw the watermarked images Guest had made and asked her to participate in a book project then in the planning stages. Gosin paired her with Bart Wasserman because of his series on the Stations of the Cross. Their choice of religious iconography and the coincidence that both artists are half Jewish and half Christian intrigued Gosin. A limited-edition book of poems by the Christian poet Czeslaw Milosz *(fig. 52)*, who was an important figure in the Polish resistance during World War II, was released by Dieu

fig. 52 Amanda Guest and Bart Wasserman, Czeslaw Milosz, *1996, 16 3/4 x 11 1/2 x 1 1/2 inches (closed)*

Donné Press in 1996. This project was a profound personal journey for Guest and extremely significant, as it was her first collaboration, first book project, and the first time she had considered an image in relation to text. For the interior illuminations, Guest made cut copper foil embedments, watermarks, stenciled pulp, and pin-pricked images. For the cover of the container for the book, Guest created a drawing of embedded nails based on the grid pattern, a motif she repeated inside the book, playing with the same pattern to reveal different layers of pulp within the paper. Susan Gosin's design of the book embodies Milosz's expression of sensual pleasure by recognizing the tactile aspects of flesh as you turn the pages and incorporating sound through the choice of different papers. For a large series of drawings executed

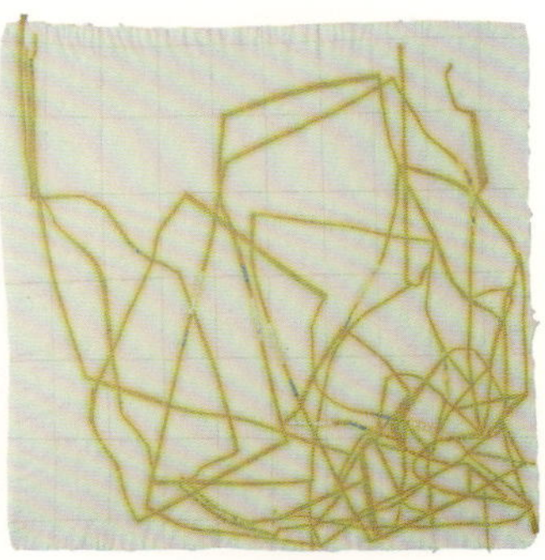

fig. 53 Amanda Guest, Yellow Bleed, *1998, 17 inches square*

over the course of a year, which includes *Yellow Bleed (fig. 53)* and *Untitled Envelope Drawing (Blue and White)* (both 1998), Guest couched a thin sheet of overbeaten translucent linen onto a felt. On top she made her drawing, composed of continuous strips cut from opened out envelopes that had enclosed correspondence from her father, an architect who lives in England. She couched another very thin linen sheet on top, embedding the drawing before it was pressed and dried. At first Guest used manila envelopes but over time her repertoire expanded. Sometimes there are snippets of handwriting or stamps from the used envelopes. The artist selected a white envelope with a blue printed inner lining for *Untitled Envelope Drawing (Blue and White)*. The title *Yellow Bleed* describes what actually happened to the drawing when it came into contact with water on the moist sheet of paper. Unlike the rectangular format of her earlier work, both of these sheets are square, which removed the vertical and horizontal orientations from the composition. Previously, Guest had been working with extremely rigid lines made of embedded nails, and these cut paper envelopes led her back to a much more gestural and freer linear style of drawing. The pieces in the exhibition are quite far along in the evolution of the series and appear more chaotic, knotted, and patterned. Subsequently, Guest started recycling earlier drawings, recutting them and using the lines to embed again. Metaphorically, the drawings are a another kind of self-portrait envelope, encapsulating memories of places the artist knew growing up.

fig. 54 Emiko Kasahara, Untitled *(memo pad and index cards), 1998, memo pad: 8 1/2 x 5 1/2 x 1/4 inches; index cards: 3 x 5 x 5/8 inches*

Emiko Kasahara's *Untitled* (memo pad and index cards) (1998) *(fig. 54)*, a collaboration with Mina Takahashi, is a trompe l'oeil masterpiece. For this multiple, a stack of thirty cards and a memo pad with twenty-five sheets are contained in a four-flap enclosure also of handmade paper and fastened by a button and string. In a maddeningly obsessive process, lines of human hair, mostly in black but some in red (for the top line of the card and the side rule of the pad) were embedded between a cotton base sheet and a thin, translucent top sheet of linen paper. Kasahara's work has a strong conceptual base. Here, paper mimics paper. She explores the borderline between art and the body by contrasting the sheer cleanliness of pristine white paper with the hair, which can be beautiful on the body, but dirty when it is dust. Referring to mass-produced paper and industrialization in general,

Kasahara's subtle homage to the *objet trouvé* is a symbol of the regulated nature of public institutions like school and office.

Since the invention and development of paper in China about A.D. 105, a Confucian ritualistic practice substitutes paper surrogates for money and necessities that are ceremonially burned to sustain the needs of the departed in the afterlife. *Burning History*, a site-specific installation presented by Paul Wong at the Neuberger Museum, Purchase, New York, in 1997, is a sweeping overview of Chinese culture and history from the perspective of ancestral worship. As a second-generation Chinese-American artist, master hand papermaker, collaborator with other artists, educator, and more recently as a researcher investigating his ancestral heritage, Wong consolidates these various identities in *Chinese Funerary Encryption* *(fig. 55)*, one element from the larger installation. This multilayered paperwork consists of appropriated images of ancient Chinese funerary objects and other Chinese cultural icons, which Wong transferred with solvent onto paper laminates made of partially burned Chinese vermilion "joss papers" embedded into handmade watermarked linen paper on a cotton base sheet. Joss paper or "spirit money" can carry printed charms and amulets used to ward off evil, procure good luck, and help the deceased buy their way through the spirit world. Because the joss paper is embedded wet, staining occurs from the water soluble red inks used to print the woodblock images. The photocopy images, which are transferred by hand, soak into the paper substrate, producing ghostly impressions. "The process of layering these images," Wong explains, "alludes to my interest in archeological excavations and the literal stratification of artifacts locked in Time."[36]

fig. 55 Paul Wong, Chinese Funerary Encryption, *1997, 30 x 40 inches*

Although Warrington Colescott created etchings to illustrate the humorous and juicy poems in *Improvisations* (1991)

fig. 56 Warrington Colescott, Improvisations *by Jerome Rothenberg, 1991, 17 5/8 x 12 x 7/8 inches (closed)*

(fig. 56) by Jerome Rothenberg, the cover was designed by Susan Gosin, employing a collage technique typical of papermaking. Because of the sexy nature of some of the poems, Colescott suggested adding condoms as well as junk found along a river's edge. A thin layer of natural flax pulp couched on top holds the collage elements in place.

For her third book project with Vincent FitzGerald & Co., Susan Weil illuminated *Giacomo Joyce* (1989) *(fig. 57)*, which replicates a little-known notebook Joyce kept when he first moved to Trieste, Italy, where he worked as a teacher. The book records erotic fantasies stimulated by Joyce's infatuation with a young student. Weil's workbooks documenting three years of intensive study of the writings of Joyce with image responses to his words were the source for the illustrations. This project represents Weil's first dialogue with Paul Wong, who developed special handmade papers using cotton and linen pulp. The collaged image of the umbrella

fig. 57 Susan Weil, Giacomo Joyce, *1989, 14 x 10 1/8 x 1 1/4 inches (closed)*

was made of pigmented linen paper dried on metal. Cutting, collage, etching, and handwork were used extensively throughout the volume. In collaboration with Wong, Weil also produced complex and lyrical paper editions for *The Warrior Ant* (1992), a poem contemplating the life cycle of the ant by writer-director Lee Breuer. Each of the thirty-five unbound pages, the front and back of which subtly connect by words and etched images, is a different color, exploring the rich possibilities of surface, color, and texture in handmade paper. Some of these editioned papers are executed with screens by embedding blue silk threads to draw lines in peach-colored paper, collaging gampi paper on Dieu Donné paper, and pulp-painting. The resulting book seamlessly merges paper, image, and text into one interdependent entity.[37]

Casting

There are two primary ways to create sculptural forms with handmade paper. In laminate casting, freshly pressed sheets of still-wet handmade paper are applied to an armature. As the paper shrinks, it conforms tightly to the frame. The more thoroughly fibers are beaten, the more they shrink. Linen, abaca, and flax are often used for this purpose because they become translucent when over-beaten. In mold casting, it is best not to use high-shrinkage pulp as it is difficult to achieve a faithful copy. Therefore, most artists use only slightly beaten cotton. Sometimes the slurry material is the consistency of thick cottage cheese and packed into a mold, or it can be made into a sheet and flopped into a mold. In either case, sponging extracts excess water and presses the pulp into the mold to get a precise impression.

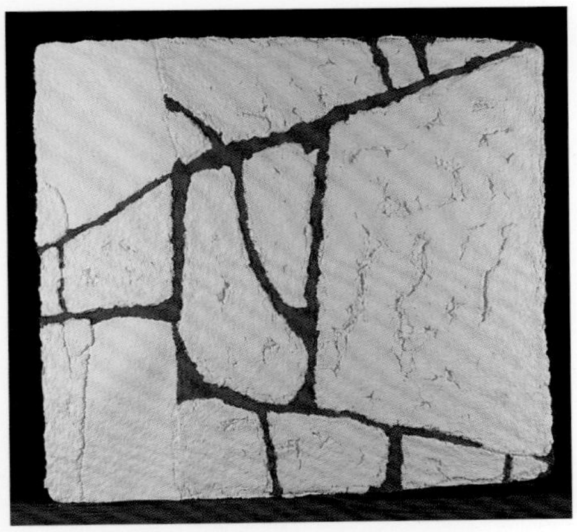

fig. 58 Golda Lewis, Cyclopean Kerb, *1981-1982, 29 3/4 x 32 inches*

Golda Lewis began experimenting with paper pulp as early as 1961, exploring the use of paper as a sculptural medium, often incorporating fragments from everyday life. Her paper wall relief *Cyclopean Kerb* (1981-82) *(fig. 58)* belongs to a series based on drawings of New York City sidewalks. On top of a pressed base sheet, the artist hand molded thick layers of cotton pulp according to a tracing on the felt. To press the work without flattening out the relief image, she placed a felt blanket over the work and a wooden board on top, then standing on the board, the artist used her body weight to flatten the pulp, retaining the relief that would be lost with a machine press. Three or more subsequent layers of pulp were added, each treated like the first with body weight flattening the pulp. The finished paperwork was slowly air-dried with fans. The dramatic fissures and abstract surface patterns in asphalt and cement pavements inspired Lewis, whose coarsely cracked sidewalk reliefs have an archaic power.

Sculptor and installation artist Winifred Lutz is also a primary innovator in the field of hand papermaking as an art form and renowned for the work she has accomplished over the past three decades. Her observation of papermaking tools and methods in Japan and Korea inspired her to combine European and Asian techniques and, through painstaking research, to invent hybrid processes that have enriched paper-casting, sheet-forming, and other methods directly related to creating art in the paper medium. Lutz has a special affinity for very thin, delicate, translucent papers. She takes advantage of the behavior of high-shrinkage pulps, selectively controlling the drying process. By restraining paper as it dries, for example, it becomes paler and lighter, while allowing it to shrink darkens the color. Lutz's palette embraces the subtle range of natural fibers, and when coloring agents are added, she prefers the reserved hues of a northern landscape. Lutz works intuitively in unconventional ways. She doesn't work from a drawing or model, but from a premise about feeling or gesture. "I discover what a piece looks like as I'm making it and that's what gives it energy."[38] It's hard to identify what the objects are because of their construction and abstract but almost familiar shapes. Lutz wants them to "have the authenticity of things and events you encounter in the natural world."[39] In some cases, her vessels are open and receptive, in others closed and secretive. For *Untitled* (1999), Lutz layered wet sheets of freshly pressed paper onto the main body of the form over a plaster mold using very overbeaten raw flax pulp. She tamped and stretched the fresh sheets tautly over the plaster, burnishing the surface and overlapping seams. When dry, paper has immense strength and reacts a little like hard leather. With one cut Lutz popped the cast off the mold. She reinforced the open end and the long narrow "beak" with 1/4-inch-thick aluminum wire coated in plastic so it wouldn't interact with the paper. She also placed a small piece of wood in the bottom or nose of the object, drilled for two pegs so it can't twist

when attached to the wall. Over the opening she stretched a very thin, freshly pressed translucent sheet of unbleached flax with superhydrated abaca squirted on top and through the mould, which watermarked the sheet like a whirlpool or the concentric rings of a tree. Light passes through the translucent paper, hits an oval of gold leaf inside the round belly of the form, and reflects back. She applied and burnished a thin strip of the same paper that she had used to cast the body of the piece to meet the edge of the top cover sheet. "Reliable vagueness is the essence of translucency," according to Lutz. "It allows light to bounce around as it passes through. Through the veil of translucency, depth and contour become ambiguous. When we peer through fog, twilight, cloudy water, our eyes experience the substance of space because we cannot see through it clearly and directly. We experience the interval. Thus, it becomes possible for the mind to be attentive to the gap between perception and definition."[40]

Some pieces take years to complete, like *A bag of pre-mix left out in the woods...* (1999) *(fig. 59)*.[41] Lutz found in the woods a discarded sack of premixed concrete, that had been partially used, dug out at the top, and left to the elements. The original paper wrapper had decayed and the concrete hardened, assuming the warm earth tones of its surroundings. Lutz carried the basin form back to her studio where it sat for a long time. When she eventually decided to use it, she carefully cast three layers of high-shrinkage unbleached line flax paper around the outside and inside of the concrete basin. She filled the interior cavity with sand and placed weights on it so the paper would shrink less and she could take advantage of the variation of color produced by the articulation of the surfaces. When the cast was dry, Lutz cut out the inside void just below the rim, which she used for another piece, and made two slits down the exterior sides, removing and reassembling the form. Then she cut a donut-shaped piece of wood the size of the basin's opening, sewed a mold for the void out of a jersey material, stapled this to the rim of the wooden

fig. 59 Winifred Lutz, A bag of pre-mix left out in the woods..., *1999,*
hanging piece: 14 x 10 x 14 inches;
floor piece: 13 x 15 x 10 inches

form, and filled it with sand. Around the suspended mold she laid a very thin laminate casting of pale to darker shades of green abaca and let it dry. By pouring the sand out, she was able to remove the dried form. Lutz also laminate-cast a wooden tree wound with opaque high-shrinkage flax. She cut a hole into the back of the laminate-cast shape and fit it with the tree-wound laminate. Depending on the light source, the aperture on the bottom allows light to enter in unexpected ways, producing a camera obscura effect. Water tinted with food dye fills the concrete basin to the limits of the applied copper leaf on the interior. The hanging element is anchored by a line of flax twine wrapped around a smooth black oval river stone inside the basin, permitting the cord to be adjusted for the height of the ceiling. Lutz is interested in the congruence and correspondence between the two forms. Each part is cast. The floating form replicates the basin to which it is anchored, yet it is not an exact copy, just as the green hues of the void mirror the water. Lutz's subtle luminescent forms compel you to look at them and try to comprehend their mysterious sensuality.

fig. 60 Arlene Shechet, Once Removed, *1998-99, 20 1/2 inches x 6 1/2 inches diameter; 15 inches x 5 1/2 inches diameter; 24 inches x 6 1/2 inches diameter*

Arlene Shechet reused some of the stenciled sheets from the *Flow Blue* series to create 100 sculptures, which she titled *Once Removed* (1998-2000) *(fig. 60)*. The first step was to have her sketches of pots translated at the wheel by a ceramist. After selecting the best shapes, she made plaster molds of them and then cast solid plaster positives of the thrown pots. Shechet covered the plaster with freshly made and lightly pressed cotton-and-stenciled abaca sheets back in the studio. She wrapped the wet sheets around the plaster forms and allowed them to shrink tightly and dry, conforming with the shape. The challenge was figuring out how to cut the paper vessel off the mold and invisibly mend it. The light paper vessels delicately balance on top of the heavy plaster forms. These sculptures illustrate Shechet's method of laminate-casting the paper as a skin or membrane and reinterpreting/reworking classic Chinese porcelains as a magnificent metaphor for crossing boundaries, growth and decay, past and present.

fig. 61 Paul Wong, Ghost Writer, *1998, 34 x 44 x 44 inches*

As part of his research into the ritualistic use of paper in Chinese culture and its relationship to ancestral worship, Paul Wong in 1998 created an installation at the Henry Street Settlement in New York, which he named *Ghost Writer (fig. 61)*, a piece triggered by the persecution of individual ideological freedoms in China, a cultural trait observed since the Qin dynasty (221 B.C.). Wong imagined a small room, a hideout or a prison cell, where a writer worked but resides no longer. The writer's possessions are concealed and embalmed with handmade abaca paper laminate-cast over found objects, taking on the role of the tortured or deceased writer's spiritual needs. Represented in conventional script and printed on the paper in the typewriter is the character for "heaven." The traditional vest hanging from the chair is covered with photocopied excerpts from the essential Confucian classics transferred to the paper, referring to the civil-service examinations scholars were forced to pass since the twelfth century to elevate their standing.

Sculptor Nancy Cohen's first experience working with paper was as a participant in the Workspace Program in 1993. She was drawn to the medium because of its fragility and strength, qualities she also admires in clay and glass. Cohen has used abaca paper as a primary material for sculpture since completing a large installation in 1995 in which garden implements and other forms were sheathed in lustrous abaca and hung in the greenhouse at the Staten Island Botanical Garden. Cohen typically forms very thin abaca sheets at the Mill, which she takes back to her studio wet and keeps refrigerated until needed. *Tussle* (1997) *(fig. 62)* was the artist's first major sculpture to use paper in many different ways within a single piece. Constructed of found materials, the understructure is carved wood and welded steel. Paper acts as a skin both technically and metaphorically. Cohen laminated abaca over a barbell-like armature, which contrasts with the pale yellow cage-like form made by stretching sheets of pigmented abaca over the steel frame and burning away the center. The objects that form the understructures of intimate assemblages like *Underbelly 2* (1999-2000) are primarily personal, domestic, and feminine (e.g., wine glasses, kitchen utensils, perfume bottles, beads). Their colors and hidden surfaces are both obscured and revealed through the transparent abaca. Cohen's enigmatic and abstract sculptures *Tussle* and *Underbelly 2* are meant to evoke a range of feelings, not one specific reading, through references to human forms, experiences, and relationships. Cohen exploits "extreme imbalances of weight and strength to provoke a direct physical reaction to the emotional content of the work."[42]

In 1992 Christy Rupp became interested in the Endangered Species Act and through her research realized it was authorized in 1972, the same year as Roe v. Wade, which compelled her to look for parallels. "I was struck," Rupp remembers, "by how the Endangered Species Act was preserving the rights of many organisms, not just the charismatic mega fauna, but also the now famous Snail Darter, insects, lichens, everything that is unique and threatened, in much the same way as Roe v. Wade sought to protect a spectrum of choice for women."[43] Rupp began to make work about absence and incompleteness. It was the beginning of her use of paper, as she wanted to integrate a skin-and-bones feeling, and paper over steel armatures is strong, even when it is torn and burned. *Moisture-Seekers* (1996/2000) *(fig. 63)* represents six frogs mutated from increasingly toxic conditions in their habitat. They are fashioned of welded steel and handmade pigmented abaca paper, which has been laminated taut over the framework and then burned. While there is never any commentary or polemic in her work, Rupp's political orientation is manifest in her beautifully constructed sculptures, which "crystallize complex social circumstances into metaphors as clear and frightening as fairy tales."[44]

fig. 62 Nancy Cohen, Tussle, *1997, 6 x 4 x 2 feet*

fig. 63 Christy Rupp, Moisture Seekers, *1996/2000, approx. 16 x 12 x 5 inches each*

Lesley Dill uses tangible materials and the body as a metaphor to investigate aspects of self and the role of women in society. During an eighteen-month stay in India in the early 1990s, Dill fell in love with paper and recognized the range of expressive possibilities it offered for her art. It was practical and easily transportable. "As a material to work with," Dill explains, "paper is the closest medium to air that I can still touch. I love its fragility and implications of violence. It rips, tears, and crumbles...but paper can always be mended. It can be sewn back together, glued, and remade."[45] A birthday gift of a book of Emily Dickinson's poetry in 1990 also had a profound effect on Dill, particularly one poem, which Dill has referenced in numerous artworks and performances for its emotional impact, including *Poem Dress (The Soul has Bandaged Moments)* (1994). Dill brought the wire sculptural framework for this work to Dieu Donné, where wet linen sheets were laminated to the delicate structure. After staining the paper with tea to signify a blend of all the peoples of the world, individually cut letters of the title of Dickinson's poem were applied with an adhesive. For Dill's ethereal 8-foot-tall *Paper Poem Dress (The Thrill came slowly like a Boon, E.D.)* (1995) *(fig. 64)*, the artist worked in stages, creating a patchwork of wire structures that were laminated in sections with linen paper. A delicately fabricated line of poetry, formed from wire dipped in pulp, becomes a string of words that drape and encircle the tall dress like an abstract thread sewing the sculpture together. Dill conceived of her dresses as skin surrogates and, like Dickinson, utilized both clothing and language to simultaneously reveal and obscure meaning.

fig. 64 Lesley Dill, Paper Poem Dress (The Thrill came slowly like a Boon, E.D.), *1995, 90 x 68 x 38 inches*

Susanna Starr was one of three artists chosen to participate in the Workspace Program in 2000. For some time she has worked exclusively with sponges, carving into them and filling them with liquid acrylic paint. Starr, whose art is about process, wanted to continue her exploration of this unusual material. Technically, her work in this exhibition, *Sponge Play (fig. 65)*, is not paper because it is made from sea sponge, which is not plant fiber. However, she employed the same methodology as working with paper. Like rag fibers, natural sea sponges were macerated into pulp. A little bit of abaca was added to this mixture to help hold it together. First, a sheet of sponge pulp was formed. Then Starr inked a hot glue drawing on plastic netting with a brayer and pressed it against the two-color sheet, printing and embossing the designs of her sculptures. After the sheet was dried she painted along the embossed line with gold, which bled green. The sheet is not soft but is porous and echoes the effect of the original material.

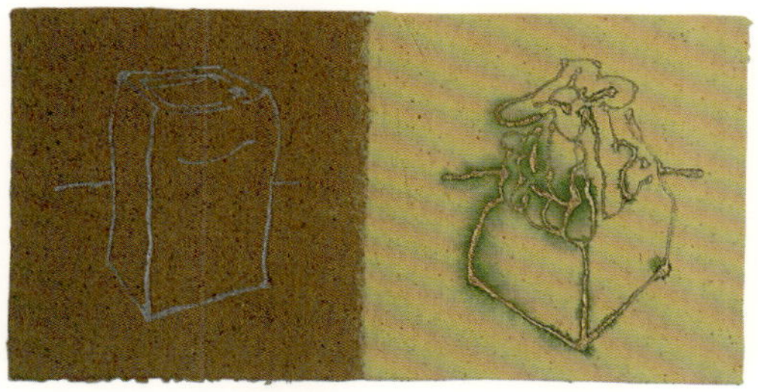

fig. 65 Susanna Starr, Sponge Play, *2000, 12 x 23 1/4 inches*

Laurie Ourlicht came to Dieu Donné to make sheets of paper to print her lithographs and fell in love with the look and feel of handmade paper. The environment of the Mill exposed her to the possibilities of what could be done with paper, and the experience, technical knowledge, and encouragement from her collaborators got her started on the path that led her out of the two-dimensional print world and into the development of cast paper objects. All of her reliefs were made by pressing paper pulp into plaster molds cast from clay forms. *Thelma and Rose* (1991) *(fig. 66)* is a loving portrait of Ourlicht's grandmothers, two strong ladies brought together by her parents' biracial marriage. Ourlicht painted the paper cast with brightly colored gouache and collaged little details to tell their stories, such as buttons to suggest Grandma Rose's work in the garment industry and the uniform to indicate Grandma Thelma's work as a domestic. The heads are dried apples carved and allowed to shrink, which create wonderful aged beings. Working in paper presented Ourlicht with more freedom to reflect on the cultural tensions of black women's evolving self-image and address questions of identity through her own personal narrative.[46]

fig. 66 Laurie Ourlicht, Thelma and Rose, *1991, 18 x 21 1/4 x 1 1/2 inches*

Walking on Eggshells (1997) *(fig. 67)*, a cibachrome photograph by Sandy Skoglund, documents a bravura work, certainly one of the most elaborate installations in the history of handmade paper as an art form. It easily has the most complex narrative of Skoglund's installations, combining two female nude models for the photograph, life-size cast paper bathroom fixtures, a floor made of thousands of empty brown eggshells, sculptures of snakes and rabbits, as well as two hieroglyphic walls of cast paper tiles with printed images of snakes, rabbits and eggs culled from the history of art and popular culture.[47] Pat Almonrode spent more than a year supervising the fabrication of this unique project, which Smith College Museum of Art commissioned for Skoglund's

fig. 67 Sandy Skoglund, Walking on Eggshells, *1997, 47 x 61 inches*

retrospective exhibition *Reality Under Siege*. Skoglund, who is known as a photographer, installation artist, painter, sculptor, and printmaker, has crafted extraordinary sculptural tableaux and documented them photographically for more than two decades. For this fantastical environment, Skoglund directed her assistants to cover a bathroom toilet, sink, mirror, and tub with plasticine, introducing an irregular texture to the surface. Rubber molds were made from the fixtures by Sculpture House in New York City and brought to Dieu Donné. Cotton linter pulp with the addition of bagasse fiber for texture was colored to match the brown chicken eggs that were to be used as the floor of the installation. Thick sheets of pulp were pressed lightly and folded into the multipart molds, then sponged by hand to conform to the shape and encourage drying. Components were cast separately, assembled and glued. In addition, some twenty-five hundred 5 x 5-inch tiles were made of the same pulp as the bathroom fixtures from forty-five different drawings by the artist. Inside a large rubber mold with four blank tile forms

fig. 68 Ming Fay, Butterfly Twig, *2001, each leaf approx. 8 inches diameter*

were placed rubber stamp images, which had been inked with a brayer. Fresh sheets were laid on top and pressed in a hydraulic press, simultaneously achieving relief printing and embossing of the graphic image and tile shape. Skoglund's artmaking strategies maintain a balance between fantasy and reality by presenting a view of life that questions what we take for granted and introducing unreal or bizarre elements in ordinary scenes. Her continuing thematic concerns – suburbia, environment, religion, dreams, food – are all manifest in *Walking on Eggshells*. Skoglund's layered and symbolic installations are planned for the vantage point of a stationary camera lens, which is no more guilty of being untruthful than anything else. "I'm thinking about the space as photographic space," explains Skoglund, "or as being a rectangle that gets filled up by these shapes. In the end that gives me a goal, which I usually overreach."[48] Nevertheless, her installations have a hybrid existence, for the artist considers individual components, such as the animal sculptures portrayed with exacting but off-kilter verisimilitude, as works of art, the installation represents the constructed environment, and the photograph is another, and in some cases, although not this one, the only lasting version of the completed dreamscape.

The continuing theme of Ming Fay's art since the early 1970s is the mimesis of nature through keen observation of plant life. "I am inspired when something unknown or ambiguous in nature catches my eye," Fay explains, "despite the fact that what is seen seems to be clear and perceivable."[49] Like Paul Wong, Fay has infused aspects of Chinese culture, philosophy, and popular mythology into his work, successfully grafting together Eastern and Western traditions. His study of Taoism, which recognizes humanity as one element in the continuum of nature, suggests it is possible through balance and change to obtain good fortune and esoteric knowledge about longevity by magical alchemy. Fay's fascination with the garden and its implications of cyclical decay and renewal have informed his still-life depictions of strangely human-scale fruit and vegetables and more recent garden archetypes. His multimedia installation *Butterfly Twig* (2001) *(fig. 68)* is based on suspension and balance. The twigs/branches defy gravity. They are made of flax pulp handmolded over a wire armature. The leaves are embedded with chamomile seeds, which are known for their specific healing properties. Their common name in Chinese herbal medicine is wooden butterfly. Formed from Chinese hemp paper that is coated with polyurethane to enhance its translucency and to give it a dewy appearance, each leaf is attached at the stem with hand-molded flax pulp. Several are grouped by twisting the stems together, and these clusters are loosely hung and balanced on the branches on site. The leaves move as air currents stir in the environment, infusing mysterious power into an enchanted natural setting.

The First Meeting of the Satie Society (1992),[50] one of the last works of art by John Cage, differs from all other artworks in the exhibition and is like no other Dieu Donné project because of the sustained collaboration required with Ben Shiff of Osiris since 1989. The first set of proofs was created between 1989 and 1991, the year before Cage died. Of the eight separate books in the valise sculpture, each illuminated by a different artist, Dieu Donné Papermill created the covers for four books and text paper for five. Because of the book form, Shiff was very concerned with the heaviness of the paper and wanted very lightweight and transparent sheets. Thus, a variety of abaca and cotton mixtures were developed. However, a high content of abaca pulp, which is susceptible to shrinkage, causes the sheets to buckle and is very sensitive to the atmosphere, subtly mimicking vellum. *Cinekus (fig. 69)*, the first in the series, for example, is completely composed of Dieu Donné paper, has an original drawing by Sol LeWitt on the cover, and contains six of his etchings. "The paper drapes almost like a fabric," Shiff observes."It is absolutely quiet, mysterious and understated."[51] Paper covers are unusual and rare in the history of book design and yet paper is vegetarian and still the best choice in terms of strength and longevity. Handmade paper is very demanding and, over time, changes and becomes stronger and more flexible, a physical correspondence with Cage's philosophy. Because of the extended time for this project and the scope and variety of different papers, Shiff has had to respond to the inherent life of each sheet he utilizes by redampening and reflattening, while exploiting drying and wrinkling as part of the imagery. All of the sheets were dried on large metal plates so the side next to the metal became crisp, shiny, luminous, buttery, and sensuous. The opposite side, where all the fiber is visible, is very chalky. By lining within the folded sheets and sewing through the open end, Shiff was able to control the arc and movement of the paper at the spine and at times subtly vary the colors within a book. Also, Shiff chose to construct some books using only one surface texture and, working from within, to create new colors by varying two differently pigmented sheets. For the twenty-four sets in this edition, the idea wasn't duplication, but to keep moving and finding opportunities to exploit the changes that were inevitable in the process.

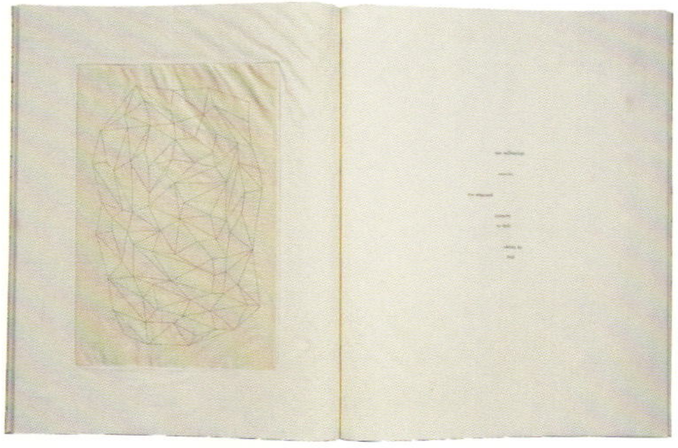

fig. 69 Sol LeWitt, Cinekus *from* The First Meeting of the Satie Society *by John Cage, 1992, 25 x 19 1/2 inches (closed)*

Rags to Riches: 25 Years of Paper Art from Dieu Donné Papermill features handmade paper as a unique medium for direct manipulation and aesthetic invention and intentionally ignores paper's more traditional function as a support for images. Dieu Donné Papermill has introduced numerous contemporary artists to this ancient handcraft. The Mill's objective has always been to expand the creative vocabulary of the artist without permitting the process to dominate. Most artists do not work exclusively in paper, so they bring their vision and knowledge to papermaking. As the beneficiaries of Dieu Donné's commitment to quality materials, technical wizardry, and unceasing innovation, the artists in the exhibition have been able to explore paper's inherent qualities, enriching their understanding and working procedures both technically and conceptually.

NOTES

1. *The author would like to thank Susan Gosin, Mina Takahashi, Paul Wong, and Pat Almonrode for their help in clarifying technical information. Also, I am extremely indebted to each of the artists who answered questions and spent time explaining the technical and conceptual basis of their art. Technical terms that appear in the essay are defined in the glossary.*

2. *Artist's written description of project, 20 September 2000.*

3. *Conversation with au hor, 17 January 2001.*

4. *Jonathan Goodman, Review,* Art in America *(January 2001): 117.*

5. *Sarah Kent,* Michelle Stuart Paperwork *(London: Institute of Contemporary Art, 1979).*

6. *E-mail from the artist, 11 April 2001.*

7. *Brett Charbeneau,* "Watermarks: Making Use of the Evidence," Hand Papermaking 8 *(winter 1993): 21.*

8. *Susan Gosin, typescript of lecture at New York University, June 1998, 16.*

9. *Margaret Mathews-Berenson,* Innovations and Explorations in Handmade Paper: Twenty Years of Collaboration at Dieu Donné Papermill *(New York: Dieu Donné Papermill, 1996), 2.*

10. *Fax from the artist, 2 February 2001.*

11. *This French metaphor, literally "spirit of the staircase," refers to a witty retort conceived after the event. For a complete discussion of this project see Susan Tallman,* "Esprit de l'Escalier: Prints by Barbara Bloom," Arts Magazine *(May 1988).*

12. *Tallman, ibid.*

13. *E-mail from the artist, 15 January 2001.*

14. *Gosin, lecture, 17.*

15. *E-mail from the artist, 3 January 2001.*

16. *A medieval French phrase, which in English means "She goes out a lot." This series of four images includes* The Studio, The Bed, The Table, *and* The Street.

17. *Artist's statement sent to author, 2 January 2001.*

18. *Timothy Barrett,* "Hand Papermaking Techniques: Western and Eastern Traditions," Making Paper *(New York: American Craft Museum, 1982), 20.*

19. *Paul Wong,* Paper as Paint *(New York: Artisan Space, Fashion Institute of Technology, 1985).*

20. *Sue Gosin,* "An Interview with Bart Wasserman," Hand Papermaking 10 *(winter 1995): 5.*

21. *Gosin, lecture, 8.*

22. *Phone conversation with author, 20 January 2001.*

23. *E-mail from the artist, 10 December 2000.*

24. *Ibid.*

25. *Artist's web site:* www.amsinet.ne.jp/~sbrayer/frame.html.

26. *E-mail from the artist, 16 January 2001.*

27. *Jonathan Goodman,* "Juan Sánchez at P.S. 1," Art in America 11 *(November 2000): 171-172.*

28. *Dore Ashton,* "Juan Sánchez," Juan Sánchez Printed Convictions: Prints and Related Works on Paper *(Jersey City, NJ: Jersey City Museum, 1999), 6.*

29. *Gosin, lecture, 13.*

30. *Jennifer Bernstein,* The Joe Wilfer Show: Collaborations in Paper and Printmaking *(Plattsburgh, NY: Plattsburgh Art Museum, State University of New York, 1998), 26.*

31. *Gosin, lecture, 13.*

32. Art On Paper *(Lund, Sweden: Anders Tornberg Gallery, 2000), 79.*

33. *The stupa form evolved from a cosmological mandala and is the architectural manifestation of the Buddha's body; it is meant to be circumambulated in a clockwise direction to acquire knowledge through experience and serenity through meditation.*

34. *Shechet's notes to author.*

35. *Carol Schwarzman,* "Margaret Lanzetta: Deborah Berke Architect's Projects New York, New York," Zingmagazine 13*: 246.*

36. *Paul Wong,* Statement, Paper Spaces *(Purchase, NY: Neuberger Museum of Art, State University of New York, 1997), 12.*

37. *Donna Stein,* Vincent FitzGerald & Company 1981-1992 *(New York: Franklin Furnace Archive, 1992);* Susan Weil, Illuminations: Bookworks of Susan Weil *(Stockholm: National Museum, 1997).*

38. *Conversation with author, 19 February 2001.*

39. *Ibid.*

40. *Artist's statement in* (Stillness) *(New York : Dieu Donné Papermill, 1999).*

41. *The full title is:*
A bag of pre-mix
Left out in the woods
Part used for who knows what.
With the rains, it hardened,
Shed its envelope,
Having taken the shape of protection
And become the color of the place.
Here, now,
Sitting upright, empty.
Brimful,
It anchors.

42. *Artist's statement in* (Stillness).

43. *E-mail from the artist, 8 January 2001.*

44. *Chris Kraus,* "Is the Cell a Happy Place?" *in* Christy Rupp: The Landscape Within *(Niagra Falls, NY: Castellani Art Museum, Niagara University, 1999).*

45. *Arlene Raven,* "Interview with Leslie Dill," Hand Papermaking 15 *(winter 2000): 3.*

46. *E-mail from the artist, 4 January 2001.*

47. *For a complete description of this project, see Linda Mueblig,* "Walking on Eggshells," Sandy Skoglund: Reality Under Siege: A Retrospective *(Northampton, MA: Smith College Museum of Art; New York: Abrams, 1998).*

48. www.artsednet.getty.edu/ArtsEDNet/hm/Mar96/0259.html.

49. "Nature Transformed," Existential Attainment *(Taipei, Taiwan: Lungmen Gallery, 1993).*

50. *The first performance of* The First Meeting of the Satie Society, *which brought form to Cage's fascination with Duchamp's concept of sculptural music, was held at the Rheinisches Landesmuseum in Bonn, Germany, on March 31, 1985. John Cage read aloud his presents for Satie. Klaus Schoening took the role of Satie and Amy Leverenz was the soprano during the Sonnekus section of the piece. See* The First Meeting of the Satie Society, *Osiris, 1994.*

51. *Conversation with publisher, 16 February 2001.*

Checklist of the Exhibition

All works courtesy of the artist unless otherwise noted. Dimensions are given in height, width, depth order.

DONALD BAECHLER

Born 1956 in Hartford, Connecticut. Lives and works in New York, New York, and Amagansett, New York.

Pompeiian Flowers from a suite of three, 1997
stenciled cotton pulp on cotton base sheet, pigment, 34 x 25 inches
published by Pace Editions, Inc.
edition of 22: PP 2/2
collaborators: Paul Wong and Ruth Lingen, Pace Editions Ink
courtesy of Dieu Donné Papermill

Three (of six) paperworks from *York House Suite*, 2000 *(fig. 42)*
stenciled linen pulp on cotton base sheet, pigment, 12 1/4 x 9 1/4 inches each
published by Pace Editions, Inc.
Edition of 50: DD
collaborators: Paul Wong and Ruth Lingen, Pace Editions Ink
assisted by Mae Shore
courtesy of Dieu Donné Papermill

LYNDA BENGLIS

Born 1941 in Lake Charles, Louisiana. Lives and works in New York, New York; East Hampton, New York; and Santa Fe, New Mexico.

Seven from the series *Game Being*, 2001
Blood Wanting, 70 inches x 7 1/2 inches diameter
Brother Animals, 70 x 15 inches x 4 inches diameter *(fig. 20)*
Earth Cavern, 42 inches x 7 inches diameter *(fig. 23)*
Flesh Devour, 53 inches x 7 1/2 inches diameter
Game Animal, 79 inches x 7 3/4 inches diameter
Prey Being, 73 x 20 inches x 8 inches diameter *(fig. 19)*
Sacred Dance, 84 inches x 7 inches diameter *(fig. 21)*
aluminum wire coils, covered with translucent abaca paper, pigment, gold-leaf flecks
collaborators: Megan Moorhouse, Lee Running, and Mina Takahashi
courtesy of the artist, Dieu Donné Papermill, and Cheim & Read, New York, New York

Sure Prey I, 2001, 16 x 19 x 5 inches
Sure Prey II, 2001, 14 x 20 x 5 inches *(fig. 22)*
Sure Prey III, 2001, 15 x 18 x 7 inches
pigmented translucent abaca paper laminate-cast from spray foam mold, gold-leafed museum board
collaborators: Megan Moorhouse, Lee Running, and Mina Takahashi
courtesy of the artist, Dieu Donné Papermill, and Cheim & Read, New York, New York

BARBARA BLOOM

Born 1951 in Los Angeles, California. Lives and works in New York, New York.

Three of seven paperworks from *Esprit de l'Escalier*, 1988 *(fig. 26)*
watermarked cotton paper, 23 1/4 x 17 3/4 inches each
published by the artist
edition of 10: unnumbered proofs
collaborator: Paul Wong
courtesy of Dieu Donné Papermill

MEL BOCHNER

Born 1940 in Pittsburgh, Pennsylvania. Lives and works in New York, New York.

Language Is Not Transparent, 1999 *(fig. 29)*
watermarked translucent abaca on pigmented cotton base sheet, 40 x 30 inches
copublished by the artist and Dieu Donné Papermill
edition of 16: PP 1/2
collaborators: Mina Takahashi and Pat Almonrode
courtesy of Mina Takahashi

SARAH BRAYER

Born 1957 in Rochester, New York. Lives and works in Kyoto, Japan, and New York, New York.

Aflame, 1988
pigmented linen pulp painting on flax base sheet, 38 x 28 1/2 inches
assisted by Paul Wong

Up Fifth Avenue, 1998 *(fig. 34)*
pigmented linen pulp painting on flax and cotton base sheet, 28 7/8 x 15 1/2 inches
assisted by Pat Almonrode

JOHN CAGE

Born 1912 in Los Angeles, California. Died 1992 in New York, New York.

Cinekus and *Words for the Satie Society* from *The First Meeting of the Satie Society*, 1992

Cinekus: pencil drawing and six etchings by Sol LeWitt on pigmented abaca and cotton paper, 25 x 19 1/2 inches (closed) *(fig. 69)*

Words for the Satie Society: pen and ink drawings by Merce Cunningham on linen paper, 22 1/2 x 13 3/4 inches (closed) and 41 1/4 inches wide (opened)

published by Osiris, New York
edition of 24: 22/24
collaborators: Paul Wong and Sharon Dürr
Courtesy of Osiris

CHUCK CLOSE

Born 1940 in Monroe, Washington. Lives and works in New York, New York.

Keith I, 1981 *(fig. 38)*
stenciled linen pulp on cotton base sheet, pigment, 35 x 26 3/4 inches
published by Pace Editions, Inc.
edition of 20: E.P.
collaborator: Joseph Wilfer
assisted by Paul Wong, Nora Shwide LeMorin and Czashka Ross
courtesy of Nora Shwide LeMorin

Self Portrait Pulp Pochoir, 2000 *(fig. 41)*
stenciled linen pulp on cotton base sheet, pigment, 24 3/4 x 19 1/2 inches
published by Pace Editions, Inc.
edition of 40: D.D. 1
collaborators: Paul Wong and Ruth Lingen, Pace Editions Ink
assisted by Mae Shore
courtesy of Dieu Donné Papermill

NANCY COHEN

Born 1959 in Queens, New York. Lives and works in Jersey City, New Jersey.

Tussle, 1997 *(fig. 62)*
abaca paper, pigmented abaca paper and flax paper over steel, wood, 6 x 4 x 2 feet
assisted by Pat Almonrode

Underbelly 2, 1999-2000
abaca paper and flax paper over wire, glass, silverware, plastic, wax, 16 x 12 x 8 inches
assisted by Pat Almonrode

WARRINGTON COLESCOTT

Born 1921 in Oakland, California. Lives and works in Hollandale, Wisconsin.

Improvisations by Jerome Rothenberg, 1991 *(fig. 56)*
etchings, letterpress, debris imbedded between flax and cotton (book cover paper), cotton paper (text papers), and linen pulp over cloth-covered case, 17 5/8 x 12 x 7/8 inches (closed)
published by Dieu Donné Press, Inc.
edition of 50: AP4
collaborator: Susan Gosin (book design, paper and letterpress)
assisted by Sharon Dürr (paper) and Paul Wong (paper and letterpress)
courtesy of Mina Takahashi and Dieu Donné Press, Inc.

EYAL DANIELI

Born in 1961 in Pittsburgh, Pennsylvania. Lives and works in New York, New York.

Bent, 1999
pigmented linen pulp painting on abaca and cotton base sheet, 24 x 18 inches
collaborator: Pat Almonrode

The Future Is Ours Comrade, 1999 *(fig. 51)*
water soluble crayon on abaca and cotton base sheet, 24 x 18 inches
collaborator: Pat Almonrode

ELENA DEL RIVERO

Born in Valencia, Spain. Lives and works in New York, New York.

The Studio (pearls) *(fig. 31)* from the series *Elle Sort Beaucoup*, 1999
mended translucent abaca paper with pearls, stitching, letterpress printing, watermarking, 60 x 40 inches
collaborators: Pat Almonrode and Lee Running
courtesy of the artist and Dieu Donné Papermill

LESLEY DILL

Born 1950 in Bronxville, New York. Lives and works in New York, New York.

Poem Dress (The Soul has Bandaged Moments), 1994
wire imbedded in linen paper with cut paper and tea stain, 31 x 22 1/2 x 1/2 inches
collaborator: Paul Wong
courtesy of Dieu Donné Papermill

Paper Poem Dress (The Thrill came slowly like a Boon, E.D.), 1995 *(fig. 64)*
wire and linen paper pulp, 90 x 68 x 38 inches
collaborator: Annie Murdock
courtesy of the artist, Dieu Donné Papermill, and George Adams Gallery, New York, New York

JIM DINE

Born 1935 in Cincinnati, Ohio. Lives and works in New York, New York.

Red Pants II, 1999 *(fig. 50)*
soft ground etching, drypoint, and hand-painting on stenciled linen pulp on cotton base sheet,
55 3/4 x 34 3/4 inches
published by Pace Editions, Inc., and Alan Cristea Galleries
edition of 20: AP II/III
collaborators: Paul Wong (paper) and Ruth Lingen (printing), Pace Editions Ink
courtesy of Dieu Donné Papermill

MELVIN EDWARDS

Born 1937 in Houston, Texas. Lives and works in New York, New York.

Everyday Connections, 2000 *(fig. 48)*
stenciled cotton/polyester rag pulp on linen base sheet, pigment, 28 1/2 x 37 1/4 inches
collaborators: Pat Almonrode and Megan Moorhouse
courtesy of Dieu Donné Papermill

Le Donné, 2000 *(fig. 49)*
stenciled cotton rag pulp on linen base sheet, pigment, 22 x 16 3/4 inches
collaborator: Pat Almonrode
courtesy of Dieu Donné Papermill

MING FAY

Born 1943 in Shanghai, China. Lives and works in New York, New York.

Butterfly Twig, 2001 *(fig. 68)*
flax pulp, Chinese hemp paper, chamomile seeds, acrylic, wire, foam, polyurethane, each leaf approximately 8 inches diameter
assisted by Paul Wong

ROBERT GOBER

Born 1954 in Wallingford, Connecticut. Lives and works in New York State.

Untitled, 2000-2001
Intaglio print on metal-dried, hand-burnished, 50% cotton linter and 50% cotton rag paper with titanium white pigment, hand torn, 2 10/16 x 3 7/16 inches
published by the artist
edition of 15: Dieu Donné Archival Proof
collaborators: Megan Moorhouse (paper) and Todd Norsten (printing)
courtesy of Dieu Donné Papermill

Untitled, 2000-2001 *(fig. 24)*
Intaglio print on hand-burnished, cotton linter paper with transparent yellow oxide and carbon black pigments, hand folded with hand-torn edges,
4 2/16 x 5 1/16 inches
published by the artist
edition of 15: Dieu Donné Archival Proof
collaborators: Megan Moorhouse (paper) and Todd Norsten (printing)
courtesy of Dieu Donné Papermill

Untitled, 2000-2001
Intaglio print on hand-burnished, 50% abaca and 50% cotton rag paper with transparent yellow oxide pigment, hand folded with hand-torn edges, 1 7/8 x 3 inches
published by the artist
edition of 15: Dieu Donné Archival Proof
collaborators: Megan Moorhouse (paper) and Todd Norsten (printing)
courtesy of Dieu Donné Papermill

SUSAN GOSIN

Born 1950 in Montclair, New Jersey. Lives and works in Princeton, New Jersey, and New York, New York.

The Lectures in Arkady by Naomi Lazard, 1986 *(figs. 27 & 28)*
etchings, letterpress, positive/negative watermarks, cotton rag (text papers), cotton, flax, and silk thread (cover paper and endsheets), 12 3/4 x 9 3/4 x 1 inches (closed)
published by Dieu Donné Press and Paper, Inc.
edition of 45: 25/45
assisted by Gail Deery and Paul Wong (paper), Leslie Myers (printing)
courtesy of Dieu Donné Press, Inc.

AMANDA GUEST

Born in 1964 in London, England. Lives and works in New York, New York.

Untitled Envelope Drawing (Blue and White), 1998
cut paper (envelope) embedded between linen paper, 17 inches square
collaborator: Paul Wong
courtesy of Dieu Donné Papermill

Yellow Bleed, 1998 *(fig. 53)*
cut paper (envelope) and thread embedded between linen paper, 17 inches square
collaborator: Paul Wong
courtesy of Dieu Donné Papermill

AMANDA GUEST AND BART WASSERMAN

Czeslaw Milosz, 1996 *(fig. 52)*
pinpricks, letterpress, watermarking and pulp painting in linen, abaca and cotton pulps, copper between linen and cotton paper (cover paper), galvanized nails between linen paper (inlaid on box-cover), linen cloth box, 16 3/4 x 11 1/2 x 1 1/2 inches (closed)
published by Dieu Donné Press, Inc.
edition of 30: 5/30
collaborators: Susan Gosin (book design and paper), Paul Wong (paper), Anne Noonan and Joe Elliot (letterpress)
assisted by Sharon Dürr (paper)
courtesy of Dieu Donné Press, Inc.

VIRGINIA JARAMILLO

Born 1939 in El Paso, Texas. Lives and works in New York, New York.

Quadrant-V, 2000 *(fig. 25)*
linen pulp painting on cotton base sheet, metallic and earth pigments, watermarked, 22 x 30 inches
collaborators: Paul Wong and Pat Almonrode

MARY JUDGE

Born in Easton, Pennsylvania. Lives and works in New York, New York.

Heaven and Earth Series #4/101, 2000 *(fig. 46)*
powdered and metallic pigment on cotton paper, embossed, 63 3/4 x 49 3/4 inches
collaborator: Paul Wong

EMIKO KASAHARA

Born 1963 in Tokyo, Japan. Lives and works in Brooklyn, New York.

Untitled (memo pad and index cards), 1998 *(fig. 54)*
human hair between linen and cotton paper, wrapped in handmade cardboard button and string enclosures; memo pad: 8 1/2 x 5 1/2 x 1/4 inches; index cards: 3 x 5 x 5/8 inches
copublished by the artist and Dieu Donné Papermill, 6/6
collaborator: Mina Takahashi
courtesy of Dieu Donné Papermill

MARGARET LANZETTA

Born 1957 in Waterbury, Connecticut. Lives and works in New York, New York.

Water Rustle II, 1999 *(fig. 47)*
pigmented cotton pulp on cotton base sheet, 30 x 22 inches
collaborator: Paul Wong
courtesy of Dieu Donné Papermill

GOLDA LEWIS

Born in Brooklyn, New York. Lives and works in New York, New York.

Cyclopean Kerb, 1981-82 *(fig. 58)*
cotton paper wall relief, 29 3/4 x 32 inches
collaborators: Paul Wong, Nora Shwide LeMorin, and Susan Gosin

WINIFRED LUTZ

Born 1942 in Brooklyn, New York. Lives and works in Huntingdon Valley, Pennsylvania.

A bag of pre-mix
Left out in the woods
Part used for who knows what.
With the rains, it hardened,
Shed its envelope,
Having taken the shape of protection
And become the color of the place.
Here, now,
Sitting upright, empty.
Brimful,
It anchors.
1999 *(fig. 59)*
cast unbleached pigmented flax, cast abaca, waxed flax thread, stone, copper leaf, dyed water, cast pre-mix concrete, cast Durham's Rock-Hard water putty; hanging piece: 14 x 10 x 14 inches; floor piece: 13 x 15 x 10 inches
collaborator: Paul Wong

Untitled, 1999
cast pigmented unbleached flax, unbleached flax painted on abaca, gold leaf, aluminum wire, 8 1/2 x 18 x 8 inches
collaborator: Paul Wong

LAURIE OURLICHT

Born 1953 in Detroit, Michigan. Lives and works in Brooklyn, New York.

Thelma and Rose, 1991 *(fig. 66)*
cast cotton pulp, gouache, cotton wadding, buttons, dried apples, mounted on Artcor,
18 x 21 1/4 x 1 1/2 inches
collaborator: Paul Wong

KEN POLINSKIE

Born 1952 in New York, New York. Lives and works in Columbia County, New York.

Still Life with Self Portrait (Forbidden Fruit), c. 1988 *(fig. 36)*
poured and painted pigmented cotton and linen pulps,
42 1/2 x 34 inches
collaborator: Paul Wong

CHRISTY RUPP

Born 1949 in Rochester, New York. Lives and works in New York, New York, and the Catskill Mountains, New York.

Six frogs from the series *Moisture-Seekers*, 1996/2000 *(fig. 63)*
steel covered with watermarked abaca paper, pigment,
variable sizes approx. 16 x 12 x 5 inches each
collaborators: Paul Wong and Pat Almonrode
assisted by Anne McKeown and David Thomson
courtesy of Dieu Donné Papermill

Two from the series *Labels for Genetically-Engineered Food*, 2000
Tell Us What We Are Eating (cucumber slices),
17 1/2 x 23 inches
Genetically Altered for Your Enjoyment (onion slices and cumin) *(fig. 30)*, 15 x 22 1/2 inches
watermarked translucent abaca paper on translucent abaca base sheet, pigment, vegetable inclusions
collaborator: Pat Almonrode
courtesy of Dieu Donné Papermill

JUAN SÁNCHEZ

Born 1954 in Brooklyn, New York. Lives and works in New York, New York, and Brooklyn, New York.

Cries and Pain, 2000 *(fig. 37)*
waterbased silkscreen, monoprint, collage and chine collé on shaped handmade cotton paper with linen pulp painting, pigment, 27 1/2 x 53 inches
copublished by the artist, Dieu Donné Papermill, and Lower East Side Printshop
edition of 10: 5/10
collaborators: Pat Almonrode and Melissa Potter (paper), Dusica Kirjakovic (printing)
courtesy of the artist, Dieu Donné Papermill, and Lower East Side Printshop

BARBARA SCHWARTZ

Born 1948 in Philadelphia, Pennsylvania. Lives and works in New York, New York.

Hexagon, 1999 *(fig. 43)*
prescotype (stenciled linen pulp on cotton base sheet, pigment), 20 1/2 x 23 inches
collaborators: Paul Wong and Pat Almonrode

ANNETTE SENNEBY

Born 1951 in Gothenburg, Sweden. Lives and works in Stockholm, Sweden.

On the Art of Painting by Jalaluddin Mahammad Rumi, 1989 *(fig. 44)*
translated by Zahra Partovi.
etchings, sculpture, letterpress, offset, metal-dried, pulp-painted, and pulp-stenciled paper (endsheets and box lining), pigmented linen (centerspread paper), 8 x 10 1/4 x 1 1/8 inches (closed)
published by Vincent FitzGerald & Co., New York
edition of 50: AP 20
collaborators: Paul Wong (paper), Marjorie Van Dyke and Vincent FitzGerald at the Printmaking Workshop, Allethaire Press, and Wild Carrot Letterpress (printing)
courtesy of Dieu Donné Papermill (gift of Dan and Gayle Weinberg) and Vincent FitzGerald & Co.

ARLENE SHECHET

Born 1951 in New York, New York. Lives and works in New York, New York.

Three vessels from *Once Removed*, 1998-99 *(fig. 60)*
stenciled pigmented linen pulp on translucent abaca base sheet, cotton paper, plaster, 20 1/2 inches x 6 1/2 inches diameter; 15 inches x 5 1/2 inches diameter; 24 inches x 6 1/2 inches diameter
collaborator: Paul Wong
assisted by Chelsea Maruskin
courtesy of the artist and Elizabeth Harris Gallery, New York, New York

Flow Blue Series: Hovering Lotus, 1999 *(fig. 45)*
stenciled linen pulp on translucent abaca base sheet, pigment, 35 x 25 inches
collaborator: Paul Wong
assisted by Dirk Staschke
courtesy of the artist and Elizabeth Harris Gallery, New York, New York

Flow Blue Series: Mirrored Sites, 2000
stenciled linen pulp on translucent abaca base sheet, pigment, 33 x 25 inches
collaborator: Paul Wong
assisted by Dirk Staschke
courtesy of the artist and Elizabeth Harris Gallery, New York, New York

ALAN SHIELDS

Born 1944 in Harrington, Kansas. Lives and works in Shelter Island, New York.

Bee Bee Wait, 2000 *(fig. 12)*
watercolor on cotton pulp-dipped galvanized steel armature, 16 x 16 x 5 inches
collaborator: Pat Almonrode
courtesy of Dieu Donné Papermill

The Bullit Train, 2000 *(fig. 13)*
watercolor on translucent abaca pulp-dipped galvanized steel armature, 15 1/2 x 15 1/2 x 3 1/2 inches

collaborators: Pat Almonrode and Lee Running
courtesy of Dieu Donné Papermill

Gate to Sweetness, 1999
watercolor on cotton pulp-dipped galvanized steel armature, 7 1/2 x 7 1/2 x 6 inches
collaborator: Anne McKeown

Me in 3D, 2000
watercolor on translucent abaca pulp-dipped galvanized steel armature, ball-bearing fishing swivel, 14 x 14 x 14 inches
collaborators: Pat Almonrode and Lee Running
courtesy of Dieu Donné Papermill

Mickie's Tools, 2000
watercolor on cotton pulp-dipped galvanized steel armature, 48 x 16 x 16 inches
collaborators: Pat Almonrode and Lee Running

Road Runner Off Ben's Point Replay, 2000 *(figs. 14 & 15)*
watercolor on cotton pulp-dipped galvanized steel armature, 23 1/2 inches diameter x 24 inches
collaborator: Pat Almonrode and Lee Running

Santa's Mousetrap, 2000
watercolor on cotton pulp-dipped galvanized steel armature, 23 1/4 x 23 1/4 x 4 3/4 inches
collaborators: Pat Almonrode and Lee Running

Sho-Gun-Long-Walk, 2000 (on view only at the Marianna Kistler Beach Museum of Art)
watercolor on abaca pulp-dipped galvanized steel armature, 16 x 16 x 7 inches
collaborators: Pat Almonrode and Lee Running
courtesy of the Sprint Corporation Art Collection, Overland Park, Kansas

The Smelly Ole' Draggon Has Teeth, 1999
watercolor on cotton pulp-dipped galvanized steel armature, 15 1/2 inches diameter x 12 inches
collaborators: Pat Almonrode and Lee Running

Stepwell, 2000 *(fig. 16)*
watercolor on cotton pulp-dipped galvanized steel armature, plexi mirror, 9 1/2 x 24 1/2 x 24 1/2 inches
collaborators: Pat Almonrode and Lee Running

SANDY SKOGLUND

Born 1946 in Boston, Massachusetts. Lives and works in New York, New York.

Walking on Eggshells, 1997 *(fig. 67)*
cibachrome photograph, 47 x 61 inches
edition of 30: 25/30

Wall tiles and cast sink for *Walking on Eggshells*, 1997
wall tiles: relief print on cotton paper with bagasse fiber inclusion, pigment, 5 1/2 x 5 1/2 x 1/8 inch each
sink: 34 1/4 x 25 1/2 x 20 3/4 inches, cast cotton paper with bagasse sugarcane fiber inclusion, pigment
collaborator: Pat Almonrode

SUSANNA STARR

Born 1960 in Ann Arbor, Michigan. Lives and works in New York, New York.

Sponge Play, 2000 *(fig. 65)*
sea sponge, abaca pulp, relief ink, marker, acrylic paint, 12 x 23 1/4 inches
collaborator: Pat Almonrode

MICHELLE STUART

Born 1940 in Los Angeles, California. Lives and works in New York, New York, and Amagansett, New York.

Zoomorph: Air, 2000 *(fig. 17)*
linen paper, wax, string,
approx. 14 x 11 inches each of twenty
collaborator: Paul Wong

Zoomorph: Earth, 2000 *(fig. 18)*
linen paper, wax, string,
approx. 14 x 11 inches each of twenty
collaborator: Paul Wong

RICHARD TUTTLE

Born 1941 in Rahway, New Jersey. Lives and works in New Mexico and New York, New York.

Hiddenness by Mei-Mei Berssenbrugge, 1987 *(fig. 35)*
etchings, relief-printing, silkscreen, letterpress, poured linen and cotton pulp images on cotton text papers, 16 1/4 x 11 x 1 1/4 inches (closed)
published by Library Fellows of the Whitney Museum of American Art, New York
edition of 120: PP
collaborator: Paul Wong
courtesy of Dieu Donné Papermill

BART WASSERMAN

Born 1930 in Philadelphia, Pennsylvania. Lives and works in Doe Run, Pennsylvania.

Veronica's Napkin, 1986 *(fig. 32)*
cotton rag pulp on linen, pressed on cotton base sheet, 30 1/4 x 22 1/4 inches
collaborators: Paul Wong and Susan Gosin

Untitled, 1999 *(fig. 33)*
cotton rag pulp on pigmented linen base sheet, 56 1/2 x 75 inches
collaborators: Paul Wong and Pat Almonrode

SUSAN WEIL

Born 1930 in New York, New York. Lives and works in New York, New York.

Giacomo Joyce by James Joyce, 1989 *(fig. 57)*
etchings, collage, letterpress, unique papers including an "umbrella" metal-dried and folded pigmented linen paper, 14 x 10 1/8 x 1 1/4 inches (closed)
published by Vincent FitzGerald & Co., New York
edition of 50: AP 20
collaborators: Paul Wong (paper), Marjorie Van Dyke and Vincent GitzGerald at the Printmaking Workshop, and Wild Carrot Letterpress (printing)
courtesy of Dieu Donné Papermill
(gift of Dan and Gayle Weinberg)

Warrior Ant by Lee Breuer, 1992
etchings, collage, silkscreen, letterpress, unique papers including a "burning eye" string-imbedded and burned cotton paper, 10 5/8 x 10 5/8 x 1 5/8 inches (closed)
published by Vincent FitzGerald & Co., New York
edition of 40: AP 10

collaborators: Paul Wong and Sharon Dürr (paper), Marjorie Van Dyke, Vincent FitzGerald, Wild Carrot Letterpress, and Colorgirls (printing)
courtesy of Dieu Donné Papermill
(gift of Dan and Gayle Weinberg)

PAUL K. WONG

Born 1951 in Fargo, North Dakota. Lives and works in New York, New York.

Chinese Funerary Encryption (fig. 55) from the installation *Burning History* at the Neuberger Museum, Purchase, NY, 1997
electrostatic transfers and collage on partially burned Chinese vermilion joss paper embedded in watermarked linen paper on cotton base sheet, 30 x 40 inches

Ghost Writer (fig. 61), from an installation at Henry Street Settlement, New York, NY, 1998
electrostatic transfers on abaca paper and abaca paper over found objects, 34 x 44 x 44 inches

Chronology

Names in bold type indicate artists included in the Rags to Riches *exhibition.*

1974

Susan Gosin and Bruce Wineberg founded Dieu Donné Press in Madison, WI • Press named after Sue's father, Dieu Donné François Gosin.

1975

Wineberg conducted initial papermaking research with commercial paper companies in the Midwest • Gosin conducted research and wrote paper on watermarking

1976

Dieu Donné Press and Paper, Inc. founded by Gosin and Wineberg, a for-profit company at 3 Crosby Street, NYC • Drafted plans for letterpress and paper studio

1977

Consulted with structural engineers and paper manufacturers • Designed, built, installed and tested papermaking equipment • Researched fiber, color, sizing • In-house training of employees • Began sheet production • Began working with **Bart Wasserman** on watermark research and art collaboration • Initiated Apprenticeship Program • First Open House Event • Gosin participated on panel at the Hand Papermaking Conference, hosted by The Center for Book Arts, NYC

1978

Received Tiffany Foundation grant for Apprenticeship Program • **Paul Wong** hired as Tiffany Apprentice • First accredited hand-papermaking class with Parsons/New School • Led papermaking workshop, Oxbow, MI • Participated in Paper, Art & Technology Conference, World Print Council Meeting, San Francisco, CA • Gosin interviewed on Voice of America

1979

Began working with **Barbara Schwartz** on sculpture edition, Barbara Gladstone publisher • Began working with **Virginia Jaramillo** on watermarked pulp paintings • Produced custom paper for **Alan Shields** • Began collaborating with Joyce Schmidt • Began collaborating with Frederic Amat • Consulted with Louis Lieberman on pulp-casting • Received NEA grant for Apprenticeship Program • Technical consultation with Joyce Schmidt for Bob Leslie Papermill, Israel

1980

Designed, printed, and produced paper for Michael Singer's *Ritual Series*, *Lilypond*, and *First Gate* • Collaborated on book edition with Naomi Spector • Collaborated with Suzanne Anker • Timothy Barrett demonstrated Japanese papermaking at Open House • Gosin moderated panel on artist collaborations, International Conference of Hand Papermakers, Boston, MA

1981

Produced first paper editions with **Chuck Close** in collaboration with Joe Wilfer, Pace Editions, Inc., publisher • Collaborated with **Golda Lewis** on cast-paper wall reliefs • Developed 100% silk rag paper for **John Cage**, *Another Song*, Callaway Editions publisher • Collaborated with Mary Frank • Produced *Li T'ai Po* from the Cathay Translation by Ezra Pound with etchings by Jane E. Goldman, Dieu Donné Press and Paper, Inc., publisher • Consulted on *Paper as Image*, touring exhibition in U.K. for Arts Council of Great Britain • Participated in exhibition *Yugoslavian-American Jubilee – 100 Years of Diplomatic Relations*, Museum of Applied Arts, Belgrade, organized by SUNY Albany, the American Embassy, and the American Agency for International Communications

1982

Began working with **Ken Polinskie** on poured-pulp paintings • Collaborated with André Kertész on portfolio • Consulted with The American Craft Museum on *Making Paper* exhibition • Consulted on exhibitions *Paper Pieces* for Freedman Gallery, Albright College, Reading, PA, and *The New Explosion: Paper Art* for the Fine Arts Museum of Long Island, Hempstead, NY • Co-organized the Fine Printing Conference, Columbia University, NYC

1983

Produced watermarked paper for **Jim Dine** for Metropolitan Opera's centennial anniversary, ULAE publisher • Began making 40 x 60-inch paper • Consulted on exhibitions *Made with Paper* for Nina Freudenheim Gallery, Buffalo, NY and *Paper: A Process Aesthetic* for Castle Gallery, New Rochelle, NY

1984

Printing studio moved off premises • Consulted on exhibition *Paper Transformed* for Turman Gallery, Indiana State University, Terre Haute, IN • Wong was visiting artist at Banff School of Art, Banff, Alberta, Canada.

1985

Researched watermarking techniques and began producing paper for *The Lectures in Arkady* by Naomi Lazard, Dieu Donné Press and Paper, Inc., publisher • Produced paper for 10 Commandments portfolio, Larry Warsh publisher • Curated *Paper and Print: Dieu Donné Artists* exhibition at Forman Gallery, Hartwick College, Oneonta, NY

1986

Began working with **Sarah Brayer** on pulp paintings • Presented *Dieu Donné Press and Paper: Book Projects* exhibition at the Watson Library, Metropolitan Museum of Art, NYC • Curated *Paper as Paint* exhibition of Dieu Donné collaborations at F.I.T., NYC • Consulted on exhibitions *Large Scale Works in Paper* for Bank of Boston Gallery, Boston, MA and *Paperworks* for Stockton State College, Pomona, NJ • Consulted on *Looking Closely* exhibition for SSC&B Lintas: Worldwide Gallery, NYC • Consulted on *Workshop Experiments* exhibition at Brattleboro Museum, VT • Served on the panel *Artists Opt for Collaboration*, organized by Pyramid Atlantic, at Corcoran School of Art, Washington, DC

1987

Dieu Donné Papermill is incorporated • Dieu Donné Press, Inc., remains imprint name of Susan Gosin's press • Collaborated with **Richard Tuttle** on book with Mei-Mei Berssenbrugge, Whitney Museum Artists and Writers Series, Whitney Museum of American Art publisher • Consulted with Robert Mapplethorpe on paper for palladium prints

1988

Dieu Donné Papermill granted not-for-profit status • Established first Board of Directors • Produced watermarked paper editions for **Barbara Bloom**, *Esprit de l'Escalier*, Venice Biennale and Hallwalls in Buffalo, NY • Produced first custom paper for publisher Vincent FitzGerald & Co. for **Susan Weil**, *The Reed* • Researched and produced bagasse (sugarcane) paper for Richard Artschwager • Produced editioned paperwork with Lucio Pozzi; Mia Feroleto and Dieu Donné copublishers

1989

Jill Bullitt hired as first Executive Director • Received first foundation grant as a nonprofit organization from the J.M. Kaplan Fund • Established Board of Advisors • Editioned paper for **Susan Weil**, *Giacomo Joyce* (James Joyce), Vincent FitzGerald & Co. publisher • Editioned paper for **Annette Senneby**, *On the Art of Painting* (Rumi), Vincent FitzGerald & Co. publisher • Designed cover and produced paper for James Merrill and Dorothea Tanning book collaboration • Sherrie Levine appropriated Dieu Donné striped paper • Editioned paper for Stephen Antonakos

1990

Mina Takahashi hired as Program Director • Workspace Program established with first project grant from the New York State Council on the Arts (NYSCA) • **Ming Fay**, **Laurie Ourlicht**, **Barbara Schwartz**, Michael Mazur inaugurated Workspace Program • Produced cover and text paper for **Sol LeWitt**, *Cinekus* from *The First Meeting of the Satie Society*, an eight-book portfolio by **John Cage** with Stephen Bastian, **Merce Cunningham**, Mell Daniel, Jasper Johns, Robert Rauschenberg, Robert Ryman, **Sol LeWitt**, Osiris publisher • Produced paper for Robert Ryman, paper monotypes, Osiris publisher • Produced cover paper for Louise Bourgeois, *The Puritan*, Osiris publisher • First issue of quarterly PULP newsletter • Organized *Handmade Paper as an Artist's/Conservator's Material* lecture series (lectures by Timothy Barrett, Laurence Barker, **Ken Polinskie**, Helen Frederick, **Winifred Lutz**, Kathryn and Howard Clark), hosted by Small Press Center, NYC, and Granary Books, NYC • Conducted high school apprenticeships through City-as-School and Star Project at Wagner

1991

Editioned paper for **Warrington Colescott**, *Improvisations*, Dieu Donné Press, Inc., publisher • Produced paper for Duane Michals, *Eros and Thanatos*, Osiris publisher • Consulted Sara Garden Armstrong on paper sculpture installation at the Bronx Museum of the Arts, NYC • Presented exhibition *Accent on Paper: 15 Years at Dieu Donné Papermill*, Lintas: Worldwide, Margaret Mathews-Berenson curator • Curated *Dieu Donné Papermill: 15 Years*, The Center for Book Arts, NYC • Consulted on exhibition *Paper-Minded: Paul Wong, Artist and Collaborator*, Pelham Art Center, Pelham, NY • **Winifred Lutz** lecture/demo • Wong lecture, *The New History*, at the New York Public Library in conjunction with exhibition

On Paper: The History of an Art (Dieu Donné collaborations represented) • Organized *Handmade Paper as an Artist's/Conservator's Material* lecture series: lectures by Joe Wilfer, Sue Allen, John Krill, and Leonard Schlosser, hosted by Cooper Union, NYC; lectures by Nancy Wu, Jared Bark, and performance by Susan Share, hosted by The Drawing Center, NYC • First membership brochures produced, Belk Mignogna designer • Launched Responsive Hands Outreach Program by Mina Takahashi and **Ken Polinskie** at Henry Street Settlement Summer Arts Program for Children, NYC

1992
Received NEA-Advancement grant for long-range planning with consultant Richard Evans • Began working with **Nancy Cohen** as Workspace Artist • Designed and produced paper for **Susan Weil**, *Warrior Ant*, Vincent FitzGerald & Co. publisher • Produced paper for Leone and Macdonald, Whitney Museum Biennial • Editioned paper for Nancy Graves print, Iris Editions publisher • Organized Conservation Endpaper Project with Library of Congress and the University of Iowa Center for the Book, Iowa City, IA • Organized lectures by Jesse Munn and Terry Belanger, hosted by Small Press Center, NYC • Organized *Paper as Ceremony* lecture series (lectures by Carol Thompson, Dorothy Field, Miguel Angel Cossio), hosted by Artists Space, NYC • Organized collaborative outreach project for children in temporary housing with El Museo del Barrio/Antonio Martorell • Assisted Japanese papermaking demonstration at Peabody & Essex Museum, Salem, MA

1993
Helen Hiebert hired as Program Director • Mina Takahashi became Executive Director • Established The Fund for Paper, Print and the Book with The Center for Book Arts to investigate shared space options • Received Arts Forward Fund grant for planning as The Fund for Paper, Print and the Book • Editioned paper for Georg Baselitz print, Limestone Press publisher • Consulted Leone and Macdonald on *Double Foolscap* project and exhibition at Whitney Museum at Philip Morris, NYC • Curated *Essential Material* book exhibition at HarperCollins, NYC • Organized *Paper Chase* event at Kate's Paperie, NYC, with The Center for Book Arts, Manhattan Graphics Center, Poets House, and The Camera Club of New York • Organized *Paper Legacies: People in the History of Hand Papermaking* lecture series (lectures by Wen Yi Hou, Elaine Koretsky, Mindell Dubansky, Bobbie Lippman), hosted by Grolier Club, NYC • Participated in Washington Art Book Fair, Washington, DC

1994
Relocation to and Phase I renovations at present 433 Broome Street location, with Norman Cox architect, Royston Barnes general contractor, and assisted by Nonprofit Facilities Fund • Joyce Mertz-Gilmore Foundation long-term viability grant • NEA-Advancement implementation grant • Copublished editions of dresses and hands with **Lesley Dill** • Copublished nail-imbedded paperworks with **Amanda Guest** • Produced celadon vase edition for Byron Kim, Whitney Museum publisher • First CEC/ArtsLink Residency with Czech artist Margita Titlova • Sponsored *Autoparticulates*, a cast paper installation by Marco Breuer and Steven Brower at the Bowling Green subway station • Presented first Dieu Donné Gallery exhibition and Grand Opening Silent Auction with art donations from more than 100 artists • Organized first benefit event and Paper Mask Ball • Opened the Mill Store • Study Collection of Handmade Paper completed by Anne Downey • Organized *Paper Legacies: People in the History of Hand Papermaking* lectures by Joanne Warner, Timothy Barrett, Cathleen Baker, Eugénie Barron, hosted by Grolier Club, NYC • Consulted Laura Anderson Barbata for her Papermaking Project by Yanomami and Ye'Kuana Indians of the Venezuelan Amazon • Served as fiscal sponsor for the Cathleen Baker/Dard Hunter Biography Project • Launched the Papermaker's Garden Project at PS 234, NYC

1995
Produced giant Kleenex tissue paper for **Robert Gober** • Collaborated with **Amanda Guest** and **Bart Wasserman** on Czeslaw Milosz book edition, Dieu Donné Press, Inc., publisher • Copublished 90-inch-tall *Paper Poem Dress* with **Lesley Dill** • Developed custom LB watermarked paper for Louise Bourgeois • *Pulp Fashion* Benefit Event and Mask Ball – 15 artists created and modeled paper fashions • *Wish You Were Here*, 1st Annual Silent Auction Postcard exhibition • Jeanne Jaffe/Leah Stein, *Dreamfield* performance in gallery • Participated in exhibition *Earth Paper* for Axis Gallery, Tokyo, Japan • Mina Takahashi, Consultant for UNIDO project in Sanganeer, India • Consultant, Handmade Cabuya Paper Project for CARE/Ecuador • Featured in *Backyard Safari*, children's television segment

1996

Began working with **Arlene Shechet** as Workspace Artist • Produced custom paper for print edition by Robert Rauschenberg, Gemini G.E.L. publisher • Organized *Symposium on Paper and Paper Art* with 38 presenters at Cooper Union • Hosted annual national meeting of the Friends of Dard Hunter at Cooper Union • Produced Dieu Donné Map of Paper Treasures in New York City • Organized *Paper Treasure Hunt*, 18 behind-the-scenes tours and studio visits • Organized *Art of the Matter* exhibition at Cooper Union (catalogue with essay by Judith K. Brodsky) • Presented *Innovations and Explorations in Handmade Paper: 20 Years of Collaboration* at the Dieu Donné Gallery, Margaret Mathews-Berenson curator (catalogue) • Invited to join the Rutgers Archives for Printmaking Studios at the Jane Voorhees Zimmerli Art Museum, New Brunswick, NJ • Featured in *Sesame Street* television segment, Joshua Selig director

1997

Pat Almonrode hired as Studio Manager • Produced cast-paper installation with **Sandy Skoglund**, *Walking on Eggshells* • Produced editioned paperwork with **Donald Baechler**, Pace Editions, Inc., publisher • Collaborated with Richard Polsky on paper tile project for Jazz Standard, NYC • Established Anita Lynn Forgach (Dhawan) Fellowship Fund • Lynn Amlie demonstrated Korean papermaking • Designed and implemented training program for Handmade Cabuya Paper Project, CARE/Ecuador, Gestamaní cloud-forest, Ecuador • Featured in WABC *New York Views* television segment

1998

NYSCA bestowed *primary institution* status to Dieu Donné • "The Crusher," 40 x 60-inch hydraulic press, arrived from Pace Editions • Copublished editioned memo pad and index cards with **Emiko Kasahara** • Collaborated with **Winifred Lutz** on NEA-funded residency • Produced custom paper for **Elena del Rivero**, *Elle Sorte Beaucoup* • Collaborated with Richard Smith on installation for British Airways Concorde Terminal at JFK Airport • Copublished editioned paperwork with Byron Kim • Hosted Greek artist Anthony Michaelides on MidAmerica Arts Alliance residency • Presented *Paper +*, Jacqueline Brody curator (catalogue) • Featured in *B. Smith with Style* television segment • Featured in *Blues Clues* television segment, Nickelodeon

1999

Stacey Lynch joined staff • Began 25th-anniversary commissions with **Alan Shields**, **Michelle Stuart**, **Lynda Benglis** • Copublished *Language Is Not Transparent* with **Mel Bochner** • Produced editioned paper for **Jim Dine**, *Red Pants II*, Pace Editions, Inc., publisher • Collaborated with **Eyal Danieli** on pulp paintings and transfer drawings • Began working with **Mary Judge** on *Heaven and Earth Series* as Workspace Artist • Began working with **Margaret Lanzetta** on stenciled pulp paintings as Workspace Artist • Began proofing with **Chuck Close** for *Self Portrait Pulp/Pochoir*, Pace Editions, Inc., publisher • Collaborated with Steve Steinman on large-scale pulp paintings • Produced custom paper for large-scale drawings by Alice Aycock • Collaborated with Sara Sosnowy on large-scale drawings • Produced archival cardboard drawing paper for Jamie Wyeth • Collaborated with Karen Davie on glass and paper project, exhibited at Mary Boone Gallery, NYC • *Drawn To Scale* exhibition at Dieu Donné and Wynn Kramarsky Space, Wynn Kramarsky curator (catalogue) • Art Publishing Research conducted by consultant Elisabeth Hahn • Dan and Gayle Weinberg donated Vincent FitzGerald & Co. publications to the Dieu Donné Archive • Searchable computerized database of Dieu Donné Archive established • Print Council of America visited • Mail-order catalog designed and illustrated by John Born • *Dieu Donné Papermaker's Cookbook* published • Organized *Dialogues in Collaboration* panel discussion on making books

2000

Megan Moorhouse joined staff • John Currie became Outreach Coordinator • Commissioned **Christy Rupp** to produce *Labels for Genetically-Engineered Food*, for *Environmentally Concerned* exhibition at Nathan Cummings Foundation, NYC • Began working with **Susanna Starr** as Workspace Artist • Copublished *Cries and Pain* with **Juan Sánchez** and Lower East Side Printshop, NYC • Collaborated with **Melvin Edwards** as the first Dieu Donné Lab Grant Artist in artist-in-residency program for midcareer artists • Began working with Jane Hammond invited as Lab Grant Artist • Began working with Dorothea Rockburne invited as Lab Grant Artist • Commissioned new work by **Elena del Rivero** for *Performance Drawings*, a collaborative project with The Drawing Center • Editioned paper for Kiki Smith, book project, Graphicstudio publisher • Consulted with Leonardo Drew on paper-casting project with The Fabric Workshop, Philadelphia, PA • Established Dieu Donné Editions Club with Robert Blackburn •

Presented first Dieu Donné Art Award to **Chuck Close**, Ruth Lingen, Richard Solomon, Joseph Wilfer, hosted by Pace Prints, NYC • American Forest and Paper Association reception, Roger Smith Hotel, NYC • Co-organized *Paper Trail* exhibition with Mason Gross School of the Arts Galleries (catalogue), Rutgers University, New Brunswick, NJ • Lobsang Samten Lama, Nima Dawa Sherpa, and Paper Road/Tibet demonstrated Himalayan papermaking

2001

Dieu Donné turned 25 • Matthew Jackson hired as Studio Manager • Erin O'Rourke hired as Administrative Assistant • Brett Littman hired as Executive Director of Administration • Produced custom paper for **Robert Gober** as part of his installation in the U.S. Pavilion, Venice Biennale • Copublished *Sky Blue Kite* with Byron Kim and Rutgers Center for Innovative Print and Paper • Editioned custom paper elements for Jonathan Lasker, Pace Editions, Inc., publisher • Commissioned works with **Elena del Rivero** and with Alison Knowles, for *Performance Drawings* series, exhibited at The Drawing Center and at the Dieu Donné Gallery • *Rags To Riches: 25 Years of Paper Art from Dieu Donné Papermill* touring exhibition premieres at Kresge Art Museum, East Lansing, Michigan, Donna Stein curator (catalogue)

History of Exhibitions

At Dieu Donné Papermill

Barbara Schwartz, *Selected Works with Dieu Donné 1979 – 1994*, Dec 94 – Jan 95

Bart Wasserman, *Recent Paperworks*, Feb – Mar 1995

Ken Polinskie, *New Works in Handmade Paper*, Mar – Apr 1995

Workspace Program: 5 Years (catalogue with essay by Laurance Wieder), Apr – Jun 1995

Lesley Dill, *An Installation*, Jun – Jul 1995

Jeanne Jaffe, *Eidetic Objects*, Sep 1995

Sarah Brayer, *Inner Essence*, Oct – Nov 1995

Kuboaa Press, *Elements*, Oct 1995

Having a Ball, Wish You Were Here: 1st Annual Silent Auction Postcard Exhibition, Nov – Dec 1995

Pulp Fashion, Nina Felshin and Elise Siegel curators, Dec 1995

Iain Machell, *Objets Perdus*, Dec 1995 – Jan 1996

Dieu Donné Editions & Series, 1988-1995, Nov 1995 – Jan 1996

Helen Frederick, *Under Construction: Relay/Rewind/Record*, Feb – Mar 1996

Thelma Mathias, *The Birthday Party*, Mar – Apr 1996

On Papermaking: A Selection of Limited Edition Books on the History and Techniques of Hand Papermaking, Mar – Apr 1996

Paul Wong, *One Billion Ghosts*, May – Jun 1996

Byron Kim, *Celadon*, Jun – Jul 1996

John Born, *Witnesses*, Jun – Jul 1996

Paper in Disguise: Dieu Donné Members' Exhibition, Aug – Sep 1996

Dan Fletcher, *Chirimen-gami: Japanese Crepe Paper*, Aug – Sep 1996

Innovations and Explorations in Handmade Paper: 20 Years of Collaboration at Dieu Donné Papermill, Margaret Mathews-Berenson curator (catalogue), Sep – Nov 1996

Joe Wilfer Remembered, Sep – Nov 1996

Reigning Cats and Dogs: 2nd Annual Silent Auction Postcard Exhibition, Nov – Dec 1996

Christy Rupp, *New Sculpture*, Dec 1996 – Feb 1997

Paperworks from the Lower East Side Printshop, curated by Sarah Lowe, Dec 1996 – Feb 1997

Sara Garden Armstrong, *Interiors Project*, Feb – Mar 1997

Ernestine Ruben, *Fusions: Paper and Photography*, Feb – Mar 1997

Virginia Jaramillo, *Black*, Apr – May 1997

Richard Hungerford, *Chicken Heads for Hands*, Jun – Jul 1997

Surface Tension: Dieu Donné Members' Exhibition, Aug – Sep 1997

Nightmare on Broome Street: 3rd Annual Silent Auction Postcard Exhibition, Oct – Nov 1997

Face to Face: Benefit Mask Exhibition, Oct – Nov 1997

Joyce McDaniel, *Holding Patterns*, Nov 1997 – Jan 1998

Laura Anderson Barbata, *In the Order of Chaos*, Jan – Feb 1998

Kirsi Tiittanen and Eija Isojarvi, *Gathering*, Feb – Apr 1998

Paper +, Jacqueline Brody curator (catalogue), Apr – May 1998

Adam Licht, *Found*, Jul – Aug 1998

Donald Baechler and Emiko Kasahara, *Recent Editions*, Sep – Oct 1998

Self-Portrait: 4th Annual Silent Auction Postcard Exhibition, Oct – Nov 1998

Dialogues in Collaboration: The Publications of Vincent FitzGerald & Co. (catalogue), Jan – Mar 1999

Bart Wasserman, *New Works*, Jun – Jul 1999

Winifred Lutz, *(stillness)* (catalogue), Mar – May 1999
Traveled to Sheehan Gallery, Whitman College, Walla Walla, WA, Jan – Feb 2000, and to Gallery Joe, Philadelphia, PA, Mar – Apr 2000

Drawn To Scale: An Exhibition of Large-Scale Works on Dieu Donné Paper (catalogue with essay by Naomi Spector), Jul – Sep 1999

Y2Kaos: 5th Annual Silent Auction Postcard Exhibition, Oct – Nov 1999

Amanda Guest, *A Given Space*, Dec 1999 – Jan 2000

Honda Takeshi, *Walking in the Mountains*, Feb – Mar 2000
Traveled in expanded version to Martin Weinstein Gallery, Minneapolis, MN, Jun – Jul 2000

Con-Text, Apr 2000

Chuck Close, *In-Progress: Self Portrait Pulp/Pochoir*, Apr – Jun 2000

Laurence Barker, Jun – Jul 2000
In conjunction with *Paper Trail*, Rutgers Summerfest Exhibition, New Brunswick, NJ (catalogue)

Golda Lewis, *Works in Paper 1974 – 1996*, Jul – Aug 2000
In conjunction with *Paper Trail*, Rutgers Summerfest Exhibition, New Brunswick, NJ (catalogue)

Alan Shields, *Intimate Details/New Paper*, Sep – Oct 2000
With concurrent exhibitions at Paula Cooper Gallery and Ned Davies Gallery, New York, NY

LandEscape: 6th Annual Silent Auction Postcard Exhibition, Oct – Nov 2000

New Editions at Dieu Donné, Nov 2000 – Jan 2001

Melvin Edwards, *Agua y Acero en Papel*, Feb – Mar 2001

Editions – RCIPP, Apr – May 2001

Alison Knowles: Paper Weather, May 2001
Part of *Performance Drawings*, a collaboration with The Drawing Center, New York (catalogue)

Douglass Morse Howell: Pioneer in Paper Art, Jun – Jul 2001

Elena del Rivero, *Las Hilanderas*, Jul 2001 (brochure); Part of *Performance Drawings*, a collaboration with The Drawing Center, New York (catalogue)

New Artist Projects at Dieu Donné, Aug 2001

For Love or Money: 7th Annual Silent Auction Postcard Exhibition, Sep – Oct 2001

Workspace Program: 1995 – 2000, Edwin Ramoran curator (catalogue), Nov 2001 – Jan 2002

At Other Venues

Paper & Print, Jan – Feb 1985
Foreman Gallery, Anderson Center for the Arts, Hartwick College, Oneonta, NY
Phil Young and Jo Mish curators

Paper as Paint, Nov – Dec 1985
Fashion Institute of Technology, New York NY
Paul Wong curator

Dieu Donné Press & Paper: Book Projects, Nov 1986
Watson Library, The Metropolitan Museum of Art, New York, NY
Susan Gosin curator

Paper-Minded: Paul Wong, Artist and Collaborator, Dec 1991
Pelham Art Center, Pelham, NY
Paul Wong curator

Dieu Donné Papermill: 15 Years, Feb – Mar 1991
The Book Arts Gallery at The Center for Book Arts, New York, NY
(catalogue)

Accent on Paper: Fifteen Years at Dieu Donné Papermill, Oct 1991 – Jan 1992
Lintas: Worldwide, New York, NY
Margaret Mathews-Berenson curator

Paper – The Essential Material: Selected Book Projects at Dieu Donné Papermill, Jan 1993
HarperCollins Publishers, New York, NY
Mina Takahashi curator (catalogue)

Handmade Paper in the Book Arts: Selected Projects at Dieu Donné Papermill, Apr 1993
St. John's University Library, Queens, NY
Traveled to Wisser Memorial Library, New York Institute of Technology, Old Westbury, NY (May – Jun 1993), Brooklyn Public Library, Brooklyn, NY (Jun – Aug 1993), and Pyramid-Atlantic, Riverdale, MD (Sep – Dec 1993)

Working the Mill: Selected Artist Collaborations at Dieu Donné Papermill, May – Nov 1993
FGIC, New York, NY
Mina Takahashi and Paul Wong curators (catalogue)

Paper, Process, Art, Feb 1994
The Art Gallery, Suffolk Community College, Selden, NY
Paul Wong curator

Essential Material: Handmade Book Projects at Dieu Donné Papermill, Aug – Oct 1994
The Berkshire Museum, Pittsfield, MA
Mina Takahashi curator (catalogue)
Traveled to Gallery II, Sangren Hall, Western Michigan University, Kalamazoo, MI (Nov 1994), and Bergstrom Mahler Museum, Neenah, WI (Dec 1994 – Feb 1995)

Marco Breuer and Steven Brower: Autoparticulates, Nov 1994 – Mar 1995
MTA Arts for Transit Exhibition Center at Bowling Green subway station, New York, NY

Dieu Donné Papermill: Selected Books and Paperworks, Feb – Mar 1995
University of Alabama Library, Tuscaloosa, AL
Mina Takahashi curator

Pulp Fusion: Recent Art from Dieu Donné Papermill, Mar – Jun 1996
5501 Columbia Art Center, Dallas, TX
Mina Takahashi curator

Green Box II: A Traveling Collection of Handmade Paper, Books and Artworks by Douglass Morse Howell, Apr – Jun 1996
Chicago Center for the Book and Paper Arts, Chicago, IL
Eugénie Barron, curator
Traveled to Brooklyn Public Library, Brooklyn, NY (Sep – Nov 1996)

Nancy Cohen: Only Connect, Jul – Nov 1996
Greenhouse Conservator Garden of the Staten Island Botanical Garden, Staten Island, NY

Art of the Matter: Seven Collaborations in Paper Art, Oct – Nov 1996
Great Hall Lobby Gallery, Cooper Union, New York, NY
(catalogue with essay by Judith K. Brodsky)
Traveled to Harry Wood Gallery, Arizona State University, Tempe, AZ
(Aug – Sep 1996)

Dard Hunter and the Art of the Handmade Book, Oct – Nov 1996
HarperCollins Publishers, New York, NY
Cathleen Baker and Dard Hunter III curators

Degrees of Presence: Paperworks from Dieu Donné Papermill, May – Oct 1997
28 North Gallery at A.T. Kearney, New York, NY
John Born curator

By Hand: The Paper of Douglass Morse Howell, May – Jul 1998
Pollock-Krasner House and Study Center, East Hampton, NY
Eugénie Barron curator

The Paper Show, Oct – Nov 1998
HBO Corporate Gallery, New York, NY
Wendy Kelley and Mina Takahashi curators

Environmentally Concerned, Apr – Sep 2000
The Nathan Cummings Foundation, New York, NY
Karen E. Jones curator (catalogue)
Traveled in expanded version to Bronx River Art Center and Gallery, Bronx, NY
(Sep – Nov 2000)

Paper Trail: Douglass Howell and Four Pioneers in American Handmade Paper: Laurence Barker, Walter Hamady, Clinton Hill, Golda Lewis, Jun – Jul 2000
Mason Gross School of the Arts Galleries, Rutgers University, The State University of New Jersey, New Brunswick, NJ
Andrea S. Honoré curator, co-organized with the Rutgers Center for Print and Paper (catalogue)

Rags To Riches: 25 Years of Paper Art from Dieu Donné Papermill, Sep – Nov 2001
premieres at Kresge Art Museum, Michigan State University, East Lansing, MI
Donna Stein curator (catalogue)
Will travel to Maryland Institute College of Art, Baltimore, MD (Nov – Dec 2001), Marianna Kistler Beach Museum of Art, Kansas State University, Manhattan, KS (Apr – Jun 2002), Heckscher Museum of Art, Huntington, Long Island, New York (Nov 2002 – Jan 2003), Milwaukee Art Museum, Milwaukee, WI (Apr – Jun 2003), and Fort Wayne Museum of Art, Fort Wayne, IN (Jul – Nov, 2003)

History of Artist Residencies

The Dieu Donné Workspace Program

1990
Ming Fay, *New York, NY*
Michael Mazur, *Boston, MA*
Laurie Ourlicht, *Brooklyn, NY*
Barbara Schwartz, *New York, NY*

1991
Tomie Arai, *New York, NY*
Carter Hodgkin, *New York, NY*
Wen Yi Hou, *New York, NY*
Kiki Smith, *New York, NY*

1992
Nancy Cohen, *Jersey City, NJ*
Michele Godwin, *New York, NY*
Timothy L. Palmer, *New York, NY*
Gloria Williams, *New York, NY*

1993
Elisa D'Arrigo, *New York, NY*
Joyce McDaniel, *Boston, MA*
Charles McQuillen, *Boston, MA*
Mary Ting, *New York, NY*

1994
Anne Chu, *Brooklyn, NY*
Gregory Coates, *New York, NY*

1995
Christine N. Blair, *Philadelphia, PA*
Matthew Lawrence, *Lancaster, PA*
Cynthia Porter, *Philadelphia, PA*
Therese Zemlin, *Boone, NC*

1996
Dale Emmart, *New York, NY*
Adam Licht, *Chicago, IL*
Shawne Major, *Highland Park, NJ*
Arlene Shechet, *New York, NY*

1997
Miriam Bloom, *New York, NY*
Lisa Corinne Davis, *New York, NY*
Cynthia Thompson, *West Hempstead, NY*

1998
Amy Bay, *New York, NY*
Margaret Lanzetta, *New York, NY*
Oona Stern, *New York, NY*

1999
Mary Judge, *New York, NY*
Jennifer Krauss, *New York, NY*
Susanna Starr, *New York, NY*

2000
Roberley A. Bell, *Batavia, NY*
Eric Hongisto, *Astoria, NY*
Karen Margolis, *New York, NY*

2001
Rosemarie Fiore, *Brooklyn, NY*
James Hegge, *Brooklyn, NY*
Marguerite Kahrl, *Brooklyn, NY*

Dieu Donné Lab Grant Program

2000
Melvin Edwards, *New York, NY*
Jane Hammond, *New York, NY*

2001
Jim Hodges, *New York, NY*
Dorothea Rockburne, *New York, NY*

International Residencies

1986
Eng Tow, *Taiwan* (Asian Cultural Council Residency)

1987
Zelio Alvis-Pinto, *Brazil* (Latin American Scholarship Program of American University)

1994
Margita Titlova, *Prague, Czech Republic* (ArtsLink/CEC Residency Program)

1996
Amalia Perjovschi, *Bucharest, Romania* (ArtsLink/CEC Residency Program)

1998
Djordje Balmazovic, *Belgrade, Serbia* (ArtsLink/CEC Residency Program)
Anthony Michaelides, *Athens, Greece* (MidAmerica Arts Alliance)

1999
Snezana Skoko Papic, *Belgrade Serbia* (ArtsLink/CEC Residency Program)

Other Residencies

Lesley Dill, *residency in studio for Dieu Donné Gallery exhibition, 1994-1995*

Christy Rupp, *residency in studio for Dieu Donné Gallery exhibition, 1996-97*

Winifred Lutz, NEA *residency, 1998-1999*

Alan Shields, *25th-anniversary commission, 1999-2000*

Amanda Guest, *six-week changing installation in Dieu Donné Gallery, 1999*

Lynda Benglis, *25th-anniversary commission, 2000-2001*

Elena del Rivero, *residency in studio for* Performance Drawings *series, collaboration with The Drawing Center, New York, NY*

Alison Knowles, *residency in studio for* Performance Drawings *series, collaboration with The Drawing Center, New York, NY*

Michelle Stuart, *25th-anniversary commission, 2000-2001*

History of Dieu Donné Editioned Works

Copublished by Dieu Donné Papermill and the artist unless otherwise noted

1975
Susan Gosin, *Fits & Burns*, edition of 18, published by Dieu Donné Press

Susan Gosin, *Prey*, edition of 25, published by Dieu Donné Press

1980
Susan Gosin, *Emily Dickinson*, edition of 20, published by Dieu Donné Press and Paper, Inc.

1981
Jane E. Goldman, *Li T'ai Po from the Cathay Translation by Ezra Pound*, edition of 25, published by Dieu Donné Press and Paper, Inc.

1986
Susan Gosin, *The Lectures in Arkady* by Naomi Lazard, edition of 45, published by Dieu Donné Press and Paper, Inc.

1988
Lucio Pozzi, *Untitled*, edition of 30, copublished by Dieu Donné Papermill and Mia Feroleto

1991
Warrington Colescott, *Improvisations* by Jerome Rothenberg, edition of 50, published by Dieu Donné Press, Inc.

1992
Irina Nakhova, *Untitled*, series of 10

1993
Lesley Dill, *Poem Dress (The Soul has Bandaged Moments)*, edition of 2

Barbara Schwartz, *Untitled*, unnumbered series

1994
Dove Bradshaw, *Indeterminance*, series of 30

Amanda Guest, *Untitled*, series of 7

Wendy Letven, *Wallpaper Compositions*, series of 10

Ken Polinskie, *Mr. & Mrs. Potatohead*, series of 12

Bart Wasserman, *Untitled*, series of 24

1995
Jackie Battenfield, *Pandora's Box*, series of 5

Jackie Battenfield, *Firecracker*, series of 6

Sarah Brayer, *Surge*, edition of 17

Carter Hodgkin, *Ozone Hole 1987*, edition of 9

1996
Amanda Guest and Bart Wasserman, *Czeslaw Milosz*, edition of 30, published by Dieu Donné Press, Inc.

1997
Katharina Hinsberg, *Still Leaves*, edition of 10

1998
Emiko Kasahara, *Untitled* (memo pad and index cards), edition of 6 each

Byron Kim, *Blue Night*, edition of 20

1999
Mel Bochner, *Language Is Not Transparent*, edition of 16

Elena del Rivero, *Elle Sort Beaucoup*, suite of 4 and unnumbered studies

Steven Steinman, *Untitled*, unnumbered series

2000
Robert Blackburn, *Sunday Afternoon*, edition of 75, published by Dieu Donné Papermill

Christy Rupp, *Moisture-Seekers*, set of 6 sculptures, edition of 2

Christy Rupp, *Labels for Genetically-Engineered Food*, unnumbered series

Juan Sánchez, *Cries and Pain*, edition of 10, copublished by the artist, Dieu Donné Papermill, and Lower East Side Printshop

Honda Takeshi, *Walking in the Mountains—February*, edition of 10

2001
Elena del Rivero, series-in-progress

Melvin Edwards, *Fragments & Shadows*, edition of 20

Byron Kim, *Sky Blue Kite*, edition of 20, copublished by the artist, Dieu Donné Papermill, and Rutgers Center for Innovative Print and Paper

Alison Knowles, *Bean Snow*, series of 30 unique prints

Jane Hammond, edition-in-progress

Paul Etienne Lincoln, edition-in-progress

Roxy Paine, edition-in-progress, copublished by Dieu Donné and Aldrich Museum of Contemporary Art

Jonathan Seliger, edition-in-progress

Artists whom we serve and who support us

THROUGH STUDIO COLLABORATIONS, CUSTOM PAPER, EXHIBITIONS, AND BENEFIT AUCTIONS

Bonnie Abbott
Lorraine Abruzzo
Suzanne Adams
Betty Alexander
Marjorie Alexander
Priscilla Algava
Jane Ingram Allen
Roberta Allen
Bethany Almonrode
Pat Almonrode
Jan Almquist
Barbara Alper
Desiree Alvarez
Frederic Amat
Peter Ambrose
Gregory Amenoff
Archer Ames
Thierry Amezcua
Grimanesa Amoros
Emma Amos
Eliana An Ghinah
William Anastasi
Craig Ancelowitz
Sue Andersen
Laura Anderson Barbata
Mary Andrews
Suzanne Anker
Alejandro Anreus
Peter Anthony
Stephen Antonakos
Pat Apatovsky
Polly Apfelbaum
Carl Apfelschnitt
Loreto Apilado
Tomie Arai
Rene Aranzamendez
Sybil Archibald
Arman
Sara Garden Armstrong
Richard Artschwager
Susan Asarian
Michael Aschenbach
Alexandra Athenassiades
Valerie Atkinson
Joana Avillez
Alice Aycock
Nancy Azara
Sabine Bachelor
John Back
Donald Baechler
Alan Baer

Cathy Baker
Lillian Ball
L.L. Balmuth
Bob Barancik
Ruth Barenbaum
Laurence Barker
Amy Barrett
Richard Barrett
Eugénie Barron
Amanda Barrow
Adam Bartos
Georg Baselitz
Leonard Baskin
Hugo Xavier Bastidas
Jackie Battenfield
Bryan Battey
Jean Battey
Amy Bay
Mark Beard
Lee Bearson
Sharon Beckman
Victoria Beerman
Lillian Bell
Roberley A. Bell
Anita Benarde
Tzvi Ben-Aretz
Sheila Benedis
Barton Benes
Lynda Benglis
Zigi Ben-Haim
Martha Benzing
Liza Berdnik
C. Rose Bergeron
Jennifer Berkun
Sarah Berney
Roger Luis Bertuzzi
Walter Biggs
Robert Blackburn
Christine Blair
Nancy Blair
Roxanne Blanchard
Bruce Bleach
Len Bloch
Jeffrey Blondes
Carmen Blondit
Barbara Bloom
Miriam Bloom
Randy Bloom
K. Bloss
Mel Bochner
Lorraine Bodger

Erika Bogin
Nina Bogin
Carol Bolt
Marcelo Bonevardi
Kevin Booth
Power Boothe
John Born
Peter Born
Andras Böröcz
Louise Bourgeois
Nancy Bowen
Jocelyn Boyea
Martin Boykan
Susan Dry Boynton
Dove Bradshaw
Maureen Brady
Sarah Brayer
Hannah Brenken
Marco Breuer
Sadie Bridger
John Brill
Robert Bringhurst
John Britton
Shannon Brock
Judith K. Brodsky
Rebecca Brooke
Bert Brouwer
Steven Brower
James Brown
Randy Brozen
Paul Bruno
Alan Bruton
Brad Buckley
Bill Bukowski
Mikaya Bullitt
June Burden
Shauna Burke
Myrna Burks
Richmond Burton
Joan Busing
Judy Byron
Catherine Cabaniss
John Cage
Mary Campbell
Richard Carboni
Carlos Cardenas
John Carnright
Kimberly Carr
Lark Carrier
Joel Carriero
Teresa Casanova

Manuel Cayupare
Julie Cencebaugh
Miguel Cervantes
Amos Chan
Grace Chan
Ann Chapman
Gerard Charrière
Jocelyn Chateauvert
Luis Chevas
Sandro Chia
Jean Chiang
Earl Childress
Nandini Chirimar
Alison Choate
Elle Chong
Terry Chouinard
Susan Chrysler White
Anne Chu
Irene Clark
Patricia Clark
Ursula Clark
Anthony Clementi
Chuck Close
Gregory Coates
Nancy Cohen
Rita Cohen
Carol Cole
Willie Cole
Ellen Coleman
Warrington Colescott
Nita Colgate
Renee Collins
William Conlon
Maureen Connor
Rona Conti
Diana Cooper
Pam Cooper
Maureen Corcoran
Chris Cordes
Alfonso Corpus
Sergi Cos-Goyon
Laura Cosgrave
Miguel Cossio
Chris Costan
Derek Coté
Laurie Coughlin
Suzan Courtney
Robert Courtright
Jennifer Crane
Danielle Creenaune
Joan Criswell

Patricia Cudd
D.D. Cummings
John Currie
Ellen Gallagher
Cynthia Gallagher
Julia D'Amario
Elisa D'Arrigo
Susan Daitch
Eyal Danieli
Carol Daniels
Nancy Davidson
Sandra Davidson
Karin Davie
Lisa Davis
Lisa Corinne Davis
Tim Davis
Georgia Deal
Gail Deery
Amanda Degener
Marc del Champ
Elena del Rivero
Pam DeLaura
Paula DeLuccia
Janice DeMarino
Amy Deneergaard
Agnes Denes
Paul Denhoed
Ann deVere
Jed Devine
Erika Devries
Jane Dickson
Lesley Dill
Kathy Dinan
Jim Dine
Mark Dion
Kathleen Doherty
Paul Donio
Seena Donneson
William Drendel
Leonardo Drew
Kelly Driscoll
Rosalyn Driscoll
Mindell Dubansky
Mona Dukess
Carroll Dunham
Aleksandar Duravcevic
Sharon Dürr
Ines Ebrenz
Jane Eccles
Marie Celeste Edwards
Melvin Edwards

Wendy Edwards
Eve Eisenstadt
Edith Ekros
Evelyn Eller
Biff Elrod
Dale Emmart
Elizabeth Enders
Engine Co. 260
Michael Ensminger
Susan Enzer
Maria Epes
Christopher Erb
Geraldine Erman
Raphael Esquer
Maria Etchepare
Doris Ettlinger
Margaret Evangeline
Donna Evans
Sue Evans
William Fares
Jane Farmer
Donald Farnsworth
Ann Farrell
Susan Fateh
Eduardo Fausti
Ming Fay
Jason Feinstein
Franklin Feldman
Walter Feldman
Annie Feldmeier
Sandi Fellman
Eileen Ferara
Dorothy Field
Rebecca Findahl
Steven Fink
Cheryl Finley
Rosemarie Fiore
Dena Paige Fischer
R. M. Fischer
Joel Fisher
George Fitzpatrick
Regina Flanagan
Dan Fletcher
L.L. Foster
Sam Francis
Ellen Frank
Mary Frank
John Franklin
Antonio Frasconi
Helen Frederick
Stephen Fredericks
Hermine Freed
Deborah Freedman
Jennifer Freeman
Lenore Freeman
Pamela Freeman
David French
Bilge Friedlander
Tova Friedman
Victor Friedman
Tavia Fromm
Gregory Frux
Leslie Fry
Hilarie Fuchs
Joan Fucillo
Jamie Fuller
John Galaskas
Phyllis Galembo
Lynn Gall
Barbara Galluci
Jo Ganter
Randy Garber
Giovanni Garcia-Fenech
Gloria Garfinkel
Julia Garretson
Deb Gates
Terasa Gejer
Sandy Gellis
Rose Generalli
Michel Gerard
Ava Gerber
John Germain
Nancy Gewirtz
Sara Gilfert
Frank Gillette
Sam Gilliam
Chris Cook Gilmore
Jake Gilson
Annika Ginsberg
Joe Ginsberg
Tatiana Ginsberg
Roland Ginzel
Joan Giordano
Candice Giove
Ronald Glassman
Robert Gober
Michele Godwin
Marsha Gold
Helen Goldberg
Maximilian Goldfarb
Jane E. Goldman
Nancy Goldring
Laurie Goldstein
Felix Gonzalez-Torres
Joe Goode
Lee Gordon
April Gornik
Karen Gorst
Susan Gosin
Lonnie Graham
Susan Graham
Nancy Graves
L.E. Green
Emma Greenberg
Kathy Greenberg
Maggie Greenwald
Jane Greer
Andrew Guenther
Amanda Guest
Christopher Guest
John Guest
Myrna Guest
Theresa Hackett
Peggy Hadden
Bill Hall
Walter Hamady
Pat Hammerman
Jane Hammond
Young Hee Han
Linda Hanauer
Lori Hansen
Ed Hanson
Rand Hardy
Elizabeth Harington
Melissa Harshman
Elizabeth Hart
Zarina Hashmi
Sarah Hauser
Elaine Hausman
Yukari Hayashida
Jim Hayes
Paula Hayes
Victoria Hayes
Courtney Healey
Stephan Heaver
Mary Heebner
James Hegge
Arnold Helbling
Neddi Heller
Pam Heller
Cathey Hemenway
Rhonda Henrich
Randy Herman
Anaida Hernández
Molly Heron
Oliver Herring
Daniel Heyman
Andrew Hibbler
Joan Hickey
Helen Hiebert
Robin Hill
Julia Hillman
Diane Himmelbaum
Katharina Hinsberg
Susan Hockaday
David Hockney
Jim Hodges
Carter Hodgkin
Judy Hoffman
Colleen Hogan
Lisa Hoke
Susan Holford
Ray Holland
Nan Hollander
Jackie Holmes
Eric Hongisto
Jon Hook
Sam Hook
Peter Hooten
Lezli Akua Hope
Brenda Horowitz
Wen Yi Hou
Dean Howell
Douglass Morse Howell
Amelia Hritz
Naomi Hudson Knapp
Rand Huebsch
Holly Hughes
David Humphrey
Jacqueline Humphries
Margo Humphries
Richard Hungerford
Sydney Hyman
Eija Isojärvi
Miriam Jacobs
Cheryl Jacobsen
Jeanne Jaffe
Lee Jaffe
Virginia Jaramillo
Jason Jasperson
David Jeffrey
Emily Jennings
Jasper Johns
Bernard Johnson
Margaret Kennard Johnson
Colette Johnson Vosberg
Mary Judge
Nori Jung
Robert Jursinski
Robin Kahn
Marguerite Kahrl
Lonny Kalfus
Mike Kamber
Sara Kamber
George Kamper
Jennifer Kapnek
Emiko Kasahara
Jak Katalan
Fran Kaufman
Megan Keary
Steven Keister
Stephen Keltner
Mary Kennedy
André Kertész
Katrin Khromin
Byron Kim
Dong Hoon Kim
Jamerry Kim
Juri Kim
Basil King
John King
Dusica Kirjakovic
Bernie Kirschenbaum
Kojino Kisakan
Brian Knoerzer
Alison Knowles
Grace Knowlton
Matthew Allen Knutzen
Ken Kobland
Dorothy Koda
Marianne Kolb
Denise Komisarek
Jane Korelitz
Kumi Korf
Fran Kornfeld
Thomas Kotik
Melanie Kozol
Jennifer Krauss
Paula Beardell Krieg
Larry Krone
Ziva Kronzon
Ahni Kruger
Melora Kuhn
Karen Kunc
Nina Kuo
Ted Kurahara
Teiko Kuroda
Janet Kusmierski
Carter Kustera
Andrei Kyryczenko
Elspeth Lamb
Bridget Lambert
M.P. Landis
Aldo Landwehr

Diana Lang
Gary Lang
Ellen Lanyon
Margaret Lanzetta
Elisa Lanzi
Jonathan Lasker
Stephanie Later
Laura Laterner
Caroline Lathan-Stiefel
Roberta Lavadour
Carla Lavatelli
Matthew Lawrence
Naomi Lazard
Sarah Leahy
Amy Lebo-Booth
Pamela D. Lee
Phyllis Lehrer
Leone & Macdonald
Ronit Leora
Marc Lepson
Marilyn Lerner
Mary Leto
Sheila Letven
Wendy Letven
Geoffrey Leven
Sherrie Levine
Barbara Levy
Ellen Levy
Leslie Lew Burns
Golda Lewis
Ross Lewis
Sol LeWitt
Josh Liberson
Adam Licht
Anna Lieberman
Louis Lieberman
Marney Lieberman
Joyce Light
Paul Etienne Lincoln
James Linehan
Ruth Lingen
Beth Lippman
Beth Littman
Mike Littman
Robert Lobe
Nora Lockshin
Betty Lofquist
Joe Loguiato
Charles Long
Edmundo Lopez
Jod Lourie
Margo Lovejoy
Jane Lozemann

Winifred Lutz
Stacey Lynch
Iain Machell
Lisa Mackie
Nicia Mafra
Shawne Major
Virginia Maksymowicz
Lenore Malen
Rallou Malliarakis
C.S. Mallinckrodt
Barry Manter
Margaret Manter
Nancy Manter
Robert Mapplethorpe
Mitchell Marco
Russell Maret
Aimée Margolis
Karen Margolis
Enid Mark
Jona Markgrat
Susan Markowitz
Allegra Marquart
Jane Marshall
Ruth Marten
Antonio Martorell
Thelma Mathias
Claudia Matzko
Michael Mazur
Peter McCabe
Louise McCagg
Maureen McCarron
Joyce McDaniel
Katie McDonnell
Dayna McErlean
Michael McFadden
Anne Q. McKeown
Margot McLean
Charles McQuillen
Sarah Mecklem
Despina Meimaroglou
Susan Melden
Sylvia Mendel
James Merrill
John Meyerhofer
Fran Meyers
Anthony Michaelides
Duane Michals
Migdal
Filipe Miguel
Diane Miller
Jessica Miller
Steve Miller
Sue Miller

Sally Minker
Richard Minsky
Patricia Miranda
Jerry Mischak
Lizbeth Mitty
Katherine Mojzsis
Brian Molanphy
Bascha Mon
Clifton Monteith
Megan Moorhouse
Roslyn Moresh
Janet Morgan
Lora Morgenstern
Laura Moriarty
R. Morosan
Elizabeth Morris
Leo Morrissey
Catherine Mosley
Cyrilla Mozenter
Paul Muldoon
Denise Mullen
Martin Mullen
Mario M. Muller
Matt Mullican
Debra Mulnick
Laurie Munn
Annie Murdock
Agnes Murray
Elizabeth Murray
Robert Murray
Peter Nadin
Irina Nakhova
James Nares
Terry Nash
Nattybumpercar
Kristin Natvig
John Nava
Joan Needham
Sunny Neeter
Walter Nelson
Vernita Nemec
Stuart Netsky
Michael Netter
Kay Newcomb
Susan Newmark
Theresa Nicholas
Carsten Nicolai
Ulf Nilsen
Takayo Noda
Tokuo Noguchi
David Novros
David Nyzio
Jennifer O'Connell

Annie O'Neill
Jennifer O'Neill
Bridgett O'Reilly
Carolyn Oberst
John Obuck
Christy Oehm
Helen Oji
Reiko Okamura
Claes Oldenburg
Richard Orient
Sherry Osorio
Martha Otis
Kristin Ott
Laurie Ourlicht
Catherine Owens
Ann Padgett
Robert Padilla
Alfonse Pagano
Jean Pagliuso
Patti Paige
Roxy Paine
Hila Palchi
Michael Palladino
Jane Palmer
Lorrie Palmer
Timothy E. Palmer
Nicholas Panayatou
Lisa Panazzolo
Flavius Panchon
Ga Hae Park
Brad Parker
Kingsley Parker
Roger Parrish
Ruth Pastine
Mariette Pathy Allen
Kaaren Patterson
Matt Pawlowski
Jeff Peachey
E.S. Pearlman
Domingo Perez
Amalia Perjovschi
John Perreault
James Perry
Andrea Peterson
Mary Peterson
Antonio Petracca
Rebecca Pike
Maria Pisano
Thomas Plageman
Felix Plaza
Bev Plummer
Lise Poirer
Lois Polansky

Anne Polashenski
Ken Polinskie
Richard Polsky
Cindy Poorbaugh
Cynthia Porter
Wayne Potratz
Melissa Potter
Joanna Pousette-D'Art
Lucio Pozzi
Jared Pratt
Mitzi Pratt
Elena Prentice
Margaret Prentice
Gail Price
Leah Raab
Michael Raab
Woody Rainey
Monisha Raja
Helen Ramsaran
Rapid Response
Robert Rauschenberg
Arianne Recto
Lowell Reiland
Fran Renda
Gail Resen
Karen Revis
Joan Rhine
Diane Rhodes
Judy Richardson
Willy Richardson
Nancy Richter
Eduardo Rincon
John Risseeuw
Paul Robbert
Annette Roberts-Gray
Sonia Robertson
Maura Robinson
Dorothea Rockburne
Eric Rodriguez
Stefan Roloff
Taney Roniger
Tori Rose
Robert Rosenbaum
Marilyn R. Rosenberg
Marc Rosenquist
Barry Rosenthal
Lorin Roser
Alan Rosner
Czashka Ross
Dene Ross
Susan Rostow
Susan Rothenberg
Barbarie Rothstein

Susan Rotolo
Susan Rowland
Ernestine Ruben
Michael Rubin
Lori Rubinfeld
Anthony Rubino
Bernette Rudolph
Cornelia Ruehlicke
Lee Emma Running
Christy Rupp
Ed Ruscha
Hugh Russell
Michael Russell
Sura Ruth
Robert Ryman
Jorge Salazar
Diane Samuels
Juan Sánchez
Jane Sangerman
Terry Savage
Claudia Sbrissa
Miriam Schaer
Lori Schafer
Kenny Scharf
Ruth Schechter
Jeffery Schiff
John Schmidt
Joyce Schmidt
Ken Schnall
Diane Schneck
Gary Schneider
Jessica Schneider
Joseph Schneider
Risa Schneider
Lynn Schnurnberger
Marian Schoettle
Ben Schonzeit
David Schorr
Linda Schrank
Joshua Schreier
Susan Schwalb
Barbara Schwartz
Gary Schwartz
Libby Seaberg
Michelle Seaman
Pat Segnan
Dewey Seid
Günther Selichar
Jonathan Seliger
Nicole Selmer
Maurice Sendak
Annette Senneby
Jinnie Seo
Richard Serra
David Shapiro
Lois Shapiro
Susan Joy Share
K. Shaw
Patricia Shea
Arlene Shechet
Trisha Shepard
Accra Shepp
Lonne Sherbill-Weinstein
Tsuneo Shibata
Alan Shields
Antje Siebrecht
Barbara Siegel
Elise Siegel
Anne Siems
Vicky and Pablo Sigwald
Laura Silver
Rosa Silver
Robbin Ami Silverberg
Ann Silverman
Bob Simon
Maggie Simonelli
M. Sinclair
Michael Singer
Skart
Sandy Skoglund
Snezana Skoko Papic
Melissa Slattery
Suzanne Slauenwhite
E.E. Smith
Joy Smith
Kiki Smith
Kimberly E. Smith
Richard Smith
Susan Smyly
David Smyth
Diana Soorikian
Joan Soppe
Sara Sosnowy
Judy Speezak
Steve Spelman
Ann Sperry
Eileen Spikol
Harley Spiller
Kira Stackhouse
Robert Stanley
Susanna Starr
Chrysanne Stathacos
Ellen Stavitsky
Steven Steinman
Sarah Stengle
Bernard Stern
Oona Stern
Judy Stevens
Caroline Stone
Todd Stone
Allyson Strafella
James Strickland
Thomas Strider
Sean Stroehle
Cindy Strunk
Patrick Strzelec
Michelle Stuart
Jean Stufflebeem
Marie Sturken
Irene Sudano
Mary Sullivan
Ilene Sunshine
Francisca Sutil
Eva Sutton
Debra Swak
Marilyn Sward
Kim Tabbat
Sara Tacheny
Mina Takahashi
Honda Takeshi
Alice Talkless
Zelda Tanenbaum
Dorothea Tanning
Janice Taylor
Joseph Taylor
T.E. Taylor
Merle Temkin
Nancy Temple
Mary Tepper
Neil Tetkowski
Christopher Thomas
Donna Thomas
Gwenn Thomas
Peter Thomas
Jamie Lou Thome
Cynthia Thompson
Kirsi Tiittanen
Mary Ting
Blaise Tobia
Patrick Todd
Fred Tomaselli
Warner Tomaszewski
Marjorie Tomchuk
Dennis Tompkins
John Tormey
Sandra Toro
Jim Torok
Vasilios A. Toulis
Eng Tow
Karen Trask
Peter Treubig
Garner Tullis
Judith Turner
Richard Tuttle
Cy Twombly
Judith Uehling
Susan Ullman
Mary Ann Unger
Tobey Urbont
Kristiina Uusitalo
Marjorie Van Dyke
Deborah Van Wetering
Gerd Vershoor
Grushenka Vicari
Jennifer Vignone
Not Vital
Gary Vogt
April Vollmer
Michael von Üchtrup
Jordi Waggoner
Merrill Wagner
Jen Wahl
M.J. Waid
Joy Walker
Reggie Walker
Lorraine Walsh
Mark Warhall
Paul Warren
Sarah Warren
Bart Wasserman
Lesley Waters
Helen Frost Way
Jeff Way
Tenesh Webber
Nancy Wechter
Bill Weege
William Wegman
Debra Weier
Susan Weil
Janna Weinstein
Margaret Weisbrod
Jessica Weiss
Judith Weiss
Cary Welch
Neil Welch
Neil Welliver
Paul West
Randy West
Steven West
Rachel Westforth
Dodi Wexler
Kit White
Beck Whitehead
Tony Whitfield
Beth Whitman
Marcia Widenor
Larry Wieder
Joseph Wilfer
Brian Williams
Gloria E. Williams
Mark Williams
Randy Williams
Tanika Williams
Christopher Wilmarth
Kate Wilson
Terry Winters
Tammy Wofsey
Muriel Wolf
Michael Wolfensohn
Eric Wong
Paul Wong
Constance Woo
Pamela Wood
T.J. Woodford
Christy Wyckoff
Jamie Wyeth
Yanomami People
Max Yawney
Ye'kuana People
Saara Ylitalo
Shoko Yomogizawa
Joanne Yoshida
Anne Youkeles
Phil Young
Ken Ytuarte
Charlotte Yudis
Susan L. Yung
Katherine Yvinskas
Stephanie Zack
Martie Zelt
Therese Zemlin
Jane Ziegler
Brenda Zlamany
Janet Zweig

Clients

Publishing

ABG Workshop
American Craft Museum
Arion Press
Art in America Book Publications
Art Institute of Chicago
Art Matters
Austin Productions
Richard Barrett
Bell Labs
Callaway Editions, Inc.
Calligraphers, Inc.
Cooper Hewitt Museum
Maureen Cummins
Derrière L'Etoile Studios
Diane Villani Editions
Dieu Donné Press, Inc.
The Drawing Center
Editions du Solstice
Elements
Daniel Elias Editions
Engine, Inc./Tokyo
Exeter Press and Paper, Inc.
The Fabric Workshop
Fashion Institute of Technology
Joe Fawbush
Mia Feroleto
Vincent FitzGerald & Co.
Flowers Graphics
Gemini G.E.L.
Barbara Gladstone Gallery
Gomez Foundation for Mill House
Graphicstudio
The Greenwall Foundation
Grenfell Press
Hand Papermaking, Inc.
Harlan & Weaver, Inc.
Jules Heller
Hine Editions, Inc.
Peter Hooten
Iris Editions
JHW Editions
Jungle Press Editions
Kaldewey Press
Alfred Knopf
Kuboaa
Limestone Press
Limited Editions Club
Nadja Press
Orion Editions
Orminda Corp.
Osiris
Pace Editions, Inc.
Pardes Rimonim Press
Jeryl Parker Editions
Perishable Press
Petersburg Press
Poison Pen Press
George Prifti, Inc.
Purgatory Pie Press
Red Butte Press at the University of Utah
Red Ozier Press
Rutgers Center for Innovative Print and Paper
Sarabande Press
Michael Singer
Sky Meadow Bindery
Soho Letterpress
Solo Impression, Inc.
Thames and Hudson
Turtle Island Press
Two Palms Press
Universal Limited Art Editions
Larry Warsh
Steve Watson
Whitney Museum of American Art Editions
Yama Prints

Conservation

American Museum of Natural History, *New York, NY*
Bill Anthony & Associates, *Chicago, IL*
Art Restoration Center, *Mahopac Falls, NY*
Boedlein Library, *Oxford, England*
BookMakers International, *Riverdale, MD*
Brigham Young University, *Provo, UT*
Brooklyn Museum of Art, *Brooklyn, NY*
Brown University, *Providence, RI*
Cleveland Museum of Art, *Cleveland, OH*
Claudia Cohen, *Easthampton, MA*
Columbia University – Butler Library, *New York, NY*
Columbia University – The Leroy Neiman Center for Print Studies, *New York, NY*
Cornell University, *Ithaca, NY*
Folger Shakespeare Library, *Washington, DC*
Greenfield Bindery, *Center Harbor, NH*
Huntington Library, *San Marino, CA*
H. P. Kraus, *New York, NY*
Intermuseum Conservation Association, *Oberlin, OH*
Library Company, *Philadelphia, PA*
Library of Congress, *Washington, DC*
William Minter, *Chicago, IL*
Martha Little Bookbinding and Conservation, *Santa Fe, NM*
Museum of Modern Art, *New York, NY*
Nancy Wu Fine Art Conservation, *New York, NY*
New York Botanical Garden Library, *Bronx, NY*
New York Historical Society, *New York, NY*
New York University – Conservation Center, Institute of Fine Arts, *New York, NY*
Newberry Library, *Chicago, IL*
Northeast Document Conservation Center, *Andover, MA*
Princeton University Firestone Library, *Princeton, NJ*
Rare Books, Inc., *Merchantville, NJ*
Rush Rhees Library, University of Rochester, *Rochester, NY*
Scalamandre, *New York, NY*
Texas Technical University Library, *Lubbock, TX*
Two Women Boxing, *Dallas, TX*
University of Alabama, *Tuscaloosa, AL*
University of Georgia – Department of Archives and History, *Athens, GA*
University of Iowa–Library Conservation Department, *Iowa City, IA*
University of Notre Dame Library, *South Bend, IN*
University of South Carolina Conservation Center, *South Park, SC*
University of Texas, *Austin, TX*
Virginia Historical Society, *Richmond, VA*
Walters Art Gallery, *Baltimore, MA*
Watson Library, Metropolitan Museum of Art, *New York, NY*
Yale Library, *New Haven, CT*

Education

Aldrich Museum of Contemporary Art, *Ridgefield, CT*
Allen Stevenson School, *New York, NY*
American Craft Museum, *New York, NY*
American Express Philanthropic Program, *New York, NY*
American Forest & Paper Association, *Washington, DC*
Art Entrée, *New York, NY*
Art Planning Consultants, *New York, NY*
Arts Connection, *New York, NY*
Auxillary Services for H.S., *New York, NY*
Banff School of Art, *Alberta, Canada*
Bank Street College of Education, *New York, NY*
The Berkshire Museum, *Pittsfield, MA*
Bridges Elementary School, *New York, NY*
Bronx River Arts Center, *Bronx, NY*
Brooklyn Children's Museum, *Brooklyn, NY*
Brooklyn Museum of Art, *Brooklyn, NY*
Caring Community Senior Center, *New York, NY*
The Center for Book Arts, *New York, NY*
CES 53, *Bronx, NY*
Chapin School, *New York, NY*
Chicago Center for Book and Paper Arts, *Chicago, IL*
Children's Discovery Center, *New York, NY*
Children's Museum of the Arts, *New York, NY*
Christie's Education, *New York, NY*
Churchill School & Center, Inc., *New York, NY*
City and Country School, *New York, NY*
City Art Tours, Inc., *New York, NY*
City-as-School, *New York, NY*
College of New Rochelle, *Bronx, NY*
Columbia University, *New York, NY*
Community Board #3 Manhattan, *New York, NY*
Congregation Beth Elohim, *Brooklyn, NY*
Cooper Hewitt Museum, *New York, NY*
Cooper Union, *New York, NY*
Corlears School, *New York, NY*
Cornell University, *Ithaca, NY*
Creative Center for Women with Cancer, *New York, NY*
Dalton School, *New York, NY*
Darmstadt Art Academy, *Darmstadt, Germany*
Discover NY with Kit Garrett, *New York, NY*
District 22, *Brooklyn, NY*
Dobbs Ferry School, *Dobbs Ferry, NY*
Dwight School, *New York, NY*
Early Childhood Center, *New York, NY*
Edward H. Bryan School, *Cresskill, NJ*
El Museo del Barrio, *New York, NY*
Enititled: Black Women Artists, *Brooklyn, NY*
Fashion Institute of Technology, *New York, NY*
Financial Women's Association of NY, *New York, NY*
Finch, Pruyn & Co. Inc., *Glens Falls, NY*
Fiorello Laguardia High School, *New York, NY*
Fordham University, *New York, NY*
Fraunces Tavern Museum, *New York, NY*
Gateway School, *New York, NY*
Gillen Brewer, *New York, NY*
Girls and Boys Club, *New York, NY*
GMHC, *New York, NY*
Gould Paper Corporation, *New York, NY*
Guggenheim/Bilbao, *Bilbao, Spain*
Guild Hall, *East Hampton, NY*
Guild of American Papercutters, *Lititz, PA*
Hartwick College, *Oneonta, NY*
Henry Street Settlement, *New York, NY*
Horticulture Society of NY, *New York, NY*
Hunter Elementary School, *New York, NY*
Irene Frank Tours, *Jericho, NY*
IS 73, *Maspeth, NY*
IS 145, *Jackson Heights, NY*
IS 218, *New York, NY*
IS 281, *Brooklyn, NY*
Jamaica Arts Center, *Jamaica, NY*
JH 210, *Ozone Park, NY*
JHS 120, *Bronx, NY*
Johnson Park School, *Princeton, NJ*
Kate's Paperie, *New York, NY*
Lafayette College, *Easton, PA*
Learning Alliance, *New York, NY*
The Learning Project, *New York, NY*
Life Center for Children, *Brooklyn, NY*
Lindenmeyer Munroe, *New York, NY*
Little Red Schoolhouse & Elizabeth Irwin High School, *New York, NY*
Long Island City High School, *Long Island City, NY*
Long Island University, *Brookville, NY*
Lou Hammond & Associates, Inc., *New York, NY*
Lower East Side Printshop, *New York, NY*
Lyceé Français, *New York, NY*
Manhattan Graphics Center, *New York, NY*
Marymount School of New York, *New York, NY*
Merritt Memorial School, *Cresskill, NJ*
Metropolitan Museum of Art, *New York, NY*
Middlesex County College, *Edison, NJ*
Milwaukee Art Museum, *Milwaukee, WI*
Mindbuilders Creative Arts Center, *Bronx, NY*
Minnesota Center for Book Arts, *Minneapolis, MN*
Museum of Modern Art, *New York, NY*
Nathan Cummings Foundation, *New York, NY*
New School for Social Research, *New York, NY*
New York Botanical Garden, *New York, NY*
New York City Alliance for Teaching Artists and Staff Development, *New York, NY*
New York City Art Teachers Association, *New York, NY*
New York Institute of Technology, Wisser Library, *Old Westbury, NY*
New York Public Library, *New York, NY*
New York University – Art Department, *New York, NY*
New York University Conservation Center, *New York, NY*
NYC Lab School, *New York, NY*
NYC Technical College, *New York, NY*
92nd Street YMCA, *New York, NY*
Old Church Cultural Center, *Demarest, NJ*
P 373S, *Staten Island, NY*
Parrish Art Museum, *Southampton, NY*
Parsons School of Art, *New York, NY*
Paul A. Kaplan Center for Educational Drama/NYU, *New York, NY*
Penland School, *Penland, NC*
Pollock-Krasner House and Study Center, *East Hampton, NY*
Poly Prep, *Brooklyn, NY*
Pratt Institute, *Brooklyn, NY*

Print Council of America, *New York, NY*
PS 2, *Jackson Heights, NY*
PS 8, *Brooklyn, NY*
PS 9, *Yonkers, NY*
PS 15, *New York, NY*
PS 29, *Brooklyn, NY*
PS 36, *Bronx, NY*
PS 38, *Brooklyn, NY*
PS 39K, *Brooklyn, NY*
PS 41, *New York, NY*
PS 69, *Jackson Heights, NY*
PS 70Q, *Long Island City, NY*
PS 78Q, *Long Island City, NY*
PS 83, *New York, NY*
PS 87, *New York, NY*
PS 91, *New York, NY*
PS 92Q, *Corona, NY*
PS 97, *New York, NY*
PS 111M, *New York, NY*
PS 112, *New York, NY*
PS 116, *New York, NY*
PS 119, *Brooklyn, NY*
PS 122, *Long Island City, NY*
PS 127, *East Elmhurst, NY*
PS 130, *Brooklyn, NY*
PS 145, *Queens, New York*
PS 146, *New York, NY*
PS 148, *East Elmhurst, NY*
PS 149, *Jackson Heights, NY*
PS 151, *Woodside, NY*
PS 166, *Long Island City, NY*
PS 169K, *Brooklyn, NY*
PS 198, *New York, NY*
PS 234, *New York, NY*
PS 251, *Brooklyn, NY*
PS 282, *Brooklyn, NY*
PS 290, *Brooklyn, NY*
PS 295K, *Brooklyn, NY*
PS 314, *Brooklyn, NY*
Retired Teachers, *Bronx, NY*
Rhode Island School of Design, *Providence, RI*
River East Elementary, *New York, NY*
Ross School, *East Hampton, NY*
Rutgers University, Mason Gross School of the Arts, *New Brunswick, NJ*
Saturday's Alive Program at Arts Connection, *New York, NY*
School of the Museum of Fine Arts, *Boston, MA*
Skidmore College, *Saratoga Springs, NY*
Snug Harbor Craft Fair, *Brooklyn, NY*
Sotheby's, *New York, NY*
Southwest School of Art and Craft, *San Antonio, TX*
Spence School, *New York, NY*
St. Augustine School, *Bronx, NY*
St. John Lutheran Church, *Richmond Hill, NY*
St. John the Baptist School, *Brooklyn, NY*
St. John's University, *Jamaica, NY*
St. Olaf College, *Northfield, MN*
Staten Island Academy, *Staten Island, NY*
Staten Island Botanical Garden, *Staten Island, NY*
State University of New York at Purchase, *Purchase, NY*
Studio in a School, *New York, NY*
Suffolk Community College, *Selden NY*
Talkart, *Rockville Centre, NY*
Trust for Public Land, *New York, NY*
UFT Teacher Center, *New York, NY*
Union College, *Schenectady, NY*
United Nations School, *Jamaica, NY*
University of Alabama at Tuscaloosa, *Tuscaloosa, AL*
University of Alabama at Birmingham, *Birmingham, AL*
University of The Arts, *Philadelphia, PA*
Virginia Commonwealth University, *Richmond, VA*
Washington Houses Community Center, *New York, NY*
Wesleyan College, *Middletown, CT*
Westchester Arts Council, *Westchester, NY*
Western Michigan University, *Kalamazoo, MI*
Whitman College, *Walla Walla, WA*
Whitney Museum of American Art, *New York, NY*
Wilson Art International, Inc., *Temple, TX*
Young Audiences, *New York, NY*

Funders

Foundations

The Arts Forward Fund
ArtsLink/CEC International Partners
Milton and Sally Avery Arts Foundation
The Mary Duke Biddle Foundation
The Buhl Foundation
Robert Sterling Clark Foundation, Inc.
The Nathan Cummings Foundation
Elizabeth Ellis Foundation
Fifth Floor Foundation
Furthermore/the publication program of
The J.M. Kaplan Fund
The Greenwall Foundation
Hardiman Family Foundation
The J.M. Kaplan Fund
Samuel H. Kress Foundation
Albert Kunstadter Family Foundation
Joyce Mertz-Gilmore Foundation
Mexican Cultural Institute of New York
Mid-America Arts Alliance
The NAN Award
The New York Community Trust
Nonprofit Finance Fund
The Starr Foundation
The Andy Warhol Foundation for the
Visual Arts, Inc.
Margaret L. Wendt Foundation
Anonymous Foundations

Corporations and Businesses

American Express Foundation
ArtCare, Inc.
Bark Frameworks, Inc.
Belk Mignogna Associates
Bookbinder's Warehouse
BookMakers International
Brooklyn Brewery
The Chase Manhattan Foundation
Chemical Bank
Citibank, N.A.
Crane & Co., Inc.
Dana Alexander Incorporated
Finch, Pruyn & Co., Inc.
Eileen Fisher, Inc.
Gould Paper Corporation
Kate's Paperie
LindenmeyerMunroe
Lou Hammand & Associates
Menasha Corporation Foundation
Neenah Printing, Menasha Corporation
New York Central Art Supply
Ocker and Trapp Library Bindery, Inc.
Pace Editions, Inc.
Simson Paper Company
Roger Smith Gallery
The Stinehour Press
Time Warner, Inc.
Victoria Paper Company, Inc.

Public Agencies

City Council Member Kathryn E. Freed,
First District, New York

City of New York,
Department of Cultural Affairs
Cultural Arts Program for Children Living in Temporary Housing
Cultural Challenge Program
Materials for the Arts
Program Services

Manhattan Community Arts Fund

National Endowment for the Arts
Advancement Program
Creation & Presentation
Planning & Stabilization
Visual Arts Program

New York State Council on the Arts
Arts in Education Program
Capital Projects
Cultural Challenge Program
Visual Arts Program

Board of Directors

Dieu Donné Press and Paper, Inc.

Susan Gosin, 1976 – 1987 *(Co-Director)*
Bruce Wineberg, 1976 – 1978 *(Co-Director)*

Dieu Donné Papermill, Inc.

Susan Gosin, 1987 – *present (President, 1987 – 2000; Co-Chair, 2001 – present)*
Richard Barrett, 1987 – 1988 *(Secretary, 1987 – 1988)*
Paul Wong, 1987 – *present (Treasurer, 1987 – 2000)*
Peter V.K. Funk, Jr., 1988 – 1993 *(Secretary, 1988 – 1993)*
Eduardo Bohórquez, 1989 – 1994
Mindell Dubansky, 1989 – 2000
Richard Minsky, 1989 – 1992
Kenneth Polinskie, 1989 – 1994
Margaret Mathews-Berenson, 1990 – 1992
Kathleen Weldon, 1990 – 1995
Joseph Wilfer, 1990 – 1994
Martin Antonetti, 1993 – 1994
Stephen Mignogna, 1993 – 1998
Elena Coon Prentice, 1993 – 1995
H. Barton Wasserman, 1993 – 1996
F. Ellen Zeifer, 1993 – 1994 *(Secretary, 1993 – 1994)*
Eric Owen Costello, 1994 – *present (Secretary, 1994 – present)*
Karen Davidson, 1995 – *present*
Gail Deery, 1995 – *present*
Katherine Hardiman, 1995 – 1997
Nancy Manter, 1995 – 1999
Audrey McGuire, 1995 – 1998
John Obuck, 1997 – 1999
James Abruzzo, 1998 – *present (Co-Chair, 2001 – present)*
Laura Stirton Aust, 1998 – *present*
Ruth Lingen, 1998 – 2000
Terry Savage, 1998 – *present*
Monica Strauss, 1998 – 2000
Rachel Fong, 1999 – *present (Treasurer, 2001 – present)*
Catherine Bindman, 2000 – *present*
Brett Littman, 2000 – 2001
Michael Wolfensohn, 2000 – *present*
Stephen Yablon, 2000 – *present*
Janice deYoung, 2001 – *present*
Margaret Pomeroy Hunt, 2001 – *present*
Cathy Lobel, 2001 – *present*
Walter Thinnes, 2001 – *present*

Honorary Directors

Eduardo Bohórquez, 1995 – *present*
Peter V.K. Funk, Jr., 1995 – *present*
Kenneth Polinskie, 1995 – *present*
Mindell Dubansky, 2001 – *present*

Board of Advisors

Laurence Barker, 1989 – *present*
Timothy Barrett, 1989 – *present*
Robert Blackburn, 1989 – *present*
Chuck Close, 1989 – *present*
Mona Dukess, 1989 – *present*
Helen Frederick, 1989 – *present*
Donald McKinney, 1989 – *present*
Elizabeth Brainard Sandwick, 1989 – 1999
Rose Slivka, 1989 – 2000
Richard Artschwager, 1990 – *present*
Barbara Lapcek, 1990 – 2000
Margaret Mathews-Berenson, 1992 – *present*
Winifred Lutz, 1995 – *present*
Katherine Hardiman, 1997 – 2000
Golda Lewis, 1998 – *present*
Jacqueline Brody, 2001 - *present*
George Negroponte, 2001 – *present*
Zelda Tanenbaum, 2001 – *present*
Kenneth Tyler, 2001 – *present*

Staff

Susan Gosin, 1976 – 1987
Andrew Ginzel, 1976 – 1978
Ray Holland, 1976 – 1978
Bruce Wineberg, 1976 – 1978
Paul Wong, 1978 – *current*
Julia Loving, 1979 – 1981
Czashka Ross, 1979 – 1980
Otavio Roth, 1981 – 1981
Nora Shwide LeMorin, 1981 – 1983
Joseph Wilfer, 1981 – 1982
Susan Anderson, 1983 – 1985
Gail Deery, 1984 – 1987
Carolyn Sheedy, 1987 – 1989
Jill Bullitt, 1989 – 1990
Sharon Dürr, 1989 – 1995
Mina Takahashi, 1990 – *current*
Helen Hiebert, 1992 – 1997
John Born, 1994 – 1998
Andrea Peterson, 1995 – 1996
Lynn Amlie, 1996 – 1997
Nita Colgate, 1996 – 1997
Cynthia Thompson, 1996 – 1996
Pat Almonrode, 1997 – 2001
Melissa Potter, 1997 – 2000
Christopher Thomas, 1998 – 1999
Stacey Lynch, 1999 – *current*
Lee Emma Running, 1999 – 2000
John Currie, 2000 – *current*
Megan Moorhouse, 2000 – *current*
Matthew Jackson, 2001 – *current*
Brett Littman, 2001 – *current*
Erin O'Rourke, 2001 – *current*
David Thomson, 2001

Instructors, Teaching Assistants, and Guest Lecturers

Pat Almonrode
Sue Allen
Lynn Amlie
Grimanesa Amoros
Susan Anderson
Jared Bark
Laurence Barker
Timothy Barrett
Eugénie Barron
Terry Belanger
Robert Blackburn
Bonnie Blitstein
John Born
Andras Borocz
Marco Breuer
Jacqueline Brody
Katherine Caraccio
May Castleberry
Kathryn Clark
Irene Clark
Howard Clark
Nita Colgate
Béatrice Coron
Miguel Angel Cosso
John Currie
Gail Deery
Amanda Degener
Kelly Driscoll
Mindell Dubansky
Sharon Dürr
Melissa Ehrenreich
Richard Evans
Jane Farmer
Donald Farnsworth
Dikko Faust
Michael Feingold
Andrea Feldman
Dorothy Field
Vincent Fitzgerald
Lynn Forgach
Helen Frederick
Mario Giacalone
Sam Gilliam
Karen Gorst
Susan Gosin
Lonnie Graham
Judith Grodowitz
Amanda Guest
Walter Hamady
Trudy Hansen
Mary Hark
Sarah Hauser
Carol Herd
Helen Hiebert
Carter Hodgkin
Judy Hoffman
Margaret Holben Ellis
Wen Yi Hou
Richard Hungerford
John Hutcheson
Virgina Jaramillo
Gunnar Kaldewey
Jim Kanter
Hiromi Katayama
Dan Keleher
Jerry Kelly
Byron Kim
Bernard Kirschenbaum
Donna Koretsky
Paula Beardell Krieg
John Krill
Lobsang Samten Lama
Ruth Lingen
Bobbie Lippman
Julia Loving
Stephanie Lussier
Winifred Lutz
Stacey Lynch
Lisa Mackie
Roberto Mannino
Abraham Marcus
Margaret Mathews-Berenson
Lee Scott McDonald
Leslie Miller
Richard Minsky
Katherine Mojzsis
Brian Molanphy
Megan Moorhouse
Laura Moriarty
Jennifer Morrow Wilson
Jessie Munn
Annie Murdoch
John Nava
Stewart Netsky
Jennifer O'Connell
Laurie Ourlicht
Zahra Partovi
Andrea Peterson
Ken Polinskie
Melissa Potter
Margaret Prentice
Katherine Radcliffe
John L. Risseeuw
Czashka Ross
Dene Ross
Otavio Roth
Lee Emma Running
Leonard Schlosser
Susan Share
Carolyn Sheedy
Nima Dawa Sherpa
Benjamin Shiff
Nora Shwide LeMorin
Robbin Ami Silverberg
E.E. Smith
Esther K. Smith
Judith Solodkin
Naomi Spector
Donna Stein
Michelle Stuart
Lynn Sures
Mina Takahashi
Zelda Tanenbaum
Christopher Thomas
Cynthia Thompson
Carol Thompson
Mary Ting
Roya Tsuchiya
Judith Turner
Carol Tyroler
April Vollmer
Suchen Wang
Joanne Warner
Jeff Way
William Weege
Susan Weil
Judith Weiss
Joseph Wilfer
Bruce Wineberg
Paul Wong
Nancy Wu
Nicholas Yeager

Interns, Work-Exchange Students, and Volunteers

Rita Alaoui
Sonia Alió
Ida Anastasia
Marina Ancona
Laura Anderson Barbata
Anne Marie Anglo
Sol Arias
Laleh Asher
Patti Banks
Jill Barber
Eugénie Barron
Anita Benarde
Laura Benzinger
Randy Bloom
John Born
Bree Branch
Nora Braverman
Jill Bressler
Marco Breuer
Laurie Brewer
Shannon Brock
Joshua Jo Brown
Loyse Brown
Sebastian Burd
Claudia Cabal
Rico Cancio
Kimberly Carr
Bob Carruthers
Ana Amelia Carvalho Luz
Piedad Ceballos
Julie Cencebaugh
Jongkee Chae
James Chase
Jocelyn Chateauvert
Arvin Chin
Cara Chinchar
Nandini Chirimar
Sook-Won Cho
Lisa Cirando
Dominique Cocuzza
Nancy Cohen
John Colella
Nita Colgate
M.J. Connors
Pam Cooper
Roberto Coromina
Michele Diana Coron
Laurie Coughlin
Jill Czarnowski
Regina D'Orazio
Carol Daniels
Saba Daraee
Kara Davis
Rachel Deitsch
Paul Denhoed
Kristen S. Denner
Serena DePero
Margaret Dessau
Ann DeVere
Tim Devin
Janice DeYoung
Carolina Dinio
Nancy Dubuar
Ngina Duckett
Celeste Edwards
Melissa Ehrenreich
Carrie Eisert
Tina Engineer
Yuko Euire
Sarah Farsad
Miriam Feinstein
Leah Finch
Cheryl Finley
Katie Fleissner
Diana Foote
Lenore Freeman
Hannah Fuller-Boswell
Laura Furmanski
John Galaskas
Lynn Gall
Matthias Geiger
Barbara Geisert
Bob George
Nuria GinÈ Gras
Tatiana Ginsberg
Annika Ginsberg
Sara Grady
Nicole Greco
Joanna Griner
Alexis Grossman
Amanda Guest
Maria Gusev
Katharine Hacklisch
Ellen Halloran
Pete Hardell
Sarah Hauser
Reva Hausman
Matt Height
Heidi Helland-Hansen
Helen Hiebert
Sarah Hill
Claire Hobden
Carter Hodgkin
Lisa Hodkoski
Carolyn Hoerle
Jaime Hoffman
Mijin Hong
Naomi Hudson-Knapp
Scott Hug
Angela Hugenschmidt
Mindy Hui
Kadiatu Ibra
Yoshiko Inaba
Miriam Jacobs
Cynthia Jimenez
Isabella Jordan
Ilsa Jule
Star Kaplan
Deborah Kaszovitz
Ted Katauskas
Betsy Katz
Erica Katz
Kathryn Kearns
Christine Kiesler
Eunice Kim
Sei Young Kim
Juri Kim
Jeun Kim
Emmie King
Maho Kino
Serena Koh
Denise M. Komisarek
Jan Kozak
Laura Kraber
Alex Krch
Lucia Kubitschek
Carter Kustera
Karen LaMonte
Christine Lawrence
Jennifer Lee
Judy Lee
Mew Ling Lee
Ing Lee
Young Jean Lee
Stacy Leeman
Richard Leibermann
Rachel Leibson
Valentina Leiderman
Jenny Lemish
Christine Lemme
Amy Lenburg
Mary Leto
Sum Li
Jill Liehler
Nora Lockshin
Nicola Lopez
John Loughlin
Katie Love
Stephanie Lussier
Ana Amelia Luz
Kathleen Madden
Adriana Magaña
Nancy Manter
Soraya Marcano
Grace Markman
Ned McBee
Dayna McErlean
Heather McGuire
Sheila McKain
Anne McKeown
Aeric Meredith-Goujon
John Meyerhofer
Jed Miner
Nobuko Misawa
Katherine Mojzsis
Brian Molanphy
Megan Moorhouse
Laura Morris
Angela Mujica
Annie Murdock
Yamini Namjoshi
Paul Narkiewicz
Maya Neumann
Bridget Nichols
Tokuo Noguchi
Ann Noonan
Susan Now
Siobhan O'Connor
Erin O'Rourke
Bob Oehl
Peggy Oei
Joanne Oliveria
Nicole Orton
Laurie Ourlicht
Jonathan Palmer
Patrick Pardo
Jason Pass
Etta Pearlman
Monique Pettway
Lida Phatthanasombati
Frank Pio
Lise Poirer
Christine Pollack
Giselle Potter
Mitzi Pratt
Blair Rainey
Monisha Raja
Jennifer Ramirez
Rick Ramos
Mary Rando
Vance Rawles
Kaitlin Redborg
Lisa Reinhold
Karen Revis
Maria Luiza Ribeiro
Malu Ribero
Jessica Richardson
Von Robinson
Jessica Rollens
Phyllis Rosenblatt
Jill Rosenfield
Dene Ross
Otavio Roth
Ernie Ruben
Laurie Rubenfeld
Lee Emma Running
Paige Sarlin
Meridith Sattler
Ruth Schechter
Risa Schneider
Tami Seymoor
Katie Shelley
Natalia Shimina
Mie Shimizu
Antje Siebrecht
Rosa Silver
Curt Smith
Kimberly Smith
Shari Smith
Michelle Snyder
Tova Sokolov
Vanina Sorrenti
Kim Sotherland
Ilya Speranza
Emily Spivak
Traian Stanescu
Marie Sturken
Irene Sudano
Farida Batool Sueda
Mariko Tada
Mina Takahashi
Zelda Tanenbaum
Irwin Tanenbaum
Ling-Yun Tang
Christopher Thomas
Cynthia Thompson
David Thomson
Charlotte Thorpe
Nuria Tomas
Ana Traversa
Linda Trumbath
Peggy Tu
Helen Tu
Jill Uehler
Samantha
VanderMeulen
Grushenka Vicari
Suchen Wang
Holly Ward
Sheila Warm
Mindy Watts
Margaret Weisbrod
Judy Weiss
Eileen Weitzman
Armando Williams
Tim Wilson
Dick Wineberg
Saya Woolfalk
Wing Y. Wu
Alyse Yang
Sunhee Yoon
Mina Young Alessandri
Julianne Zaleta
Talin Zerdelian
Jenny Zhang
Peter Zuratynsky

Selected Bibliography

Publications in each section are presented chronologically.

Dieu Donné Publications

PULP Newsletter. *New York: Dieu Donné Papermill, 1991 - present.*

essential material: handmade book projects. *New York: Dieu Donné Papermill, 1994. Introduction by Mina Takahashi.*

Workspace Program: 5 Years. *New York: Dieu Donné Papermill, 1995. Introduction by Mina Takahashi. Essay by Laurence Wieder.*

Innovations and Explorations in Handmade Paper: Twenty Years of Collaborations at Dieu Donné Papermill. *New York: Dieu Donné Papermill, 1996. Introduction by Susan Gosin. Essay by Margaret Mathews Berenson, curator.*

Dieu Donné Map of Paper Treasures in New York City. *New York: Dieu Donné Papermill, 1996.*

Paper +: Works on Dieu Donné Paper. *New York: Dieu Donné Papermill, 1998. Introduction by Susan Gosin. Essay by Jacqueline Brody, curator.*

Dialogues in Collaboration: The Publications of Vincent FitzGerald & Co. *New York: Dieu Donné Papermill, 1999. Introduction by Paul Wong. Essay by Michael Feingold.*

Drawn to Scale: An Exhibition of Large-Scale Works on Dieu Donné Paper. *New York: Dieu Donné Papermill, 1999. Introduction by Mina Takahashi. Essay by Naomi Spector.*

Potter, Melissa. Editor. Dieu Donné Hand Papermakers' Cookbook. *New York: Dieu Donné Papermill, 1999.*

Winifred Lutz: (stillness). *New York: Dieu Donné Papermill, 1999. Introduction by Mina Takahashi. Essay by Janet Koplos.*

Environmentally Concerned. *New York: Dieu Donné Papermill, 2000. Essay by Karen E. Jones, curator.*

Melvin Edwards: Agua y Acero en Papel. *Dieu Donné Lab Grant Program Publication Series, no.1. New York: Dieu Donné Papermill, 2001. Introduction by Mina Takahashi. Conversation between Melvin Edwards and Pat Almonrode, moderated by Mina Takahashi.*

Elena del Rivero: Las Hilanderas. *New York: Dieu Donné Papermill, 2001. Essay by Mina Takahashi*

Workspace Program: 1995 – 2000 *(in production). New York: Dieu Donné Papermill, 2001. Introduction by Mina Takahashi. Essay by Edwin Ramoran, curator.*

Books and Exhibition Catalogues

Heller, Jules. Papermaking. *New York: Watson-Guptill Publications, 1978; 170.*

Askin, Lucille Bunin. 26th Annual Contemporary American Art Exhibition: 1981 Paperworks. *Bethlehem, PA: Ralph Wilson Gallery, Lehigh University, 1981.*

Gosin, Sue. "Artist Collaborations." *In* International Conference of Hand Papermakers. *Boston: Carriage House Press, 1981; 52-57.*

Making Paper: Papermaking USA History, Process, Art and The Handmade Paper Book. *New York: American Craft Museum, 1982; 3, 21, 23, 49-53, 60-62.*

Schmidt, Joyce and Tova Tsuriel. Through the Mill. *Tel Aviv: American Cultural Center, 1982.*

Flomenhaft, Eleanor. The New Explosion: Paper Art. *New York: CDS Gallery, New York, 1983.*

Proceedings of The Fine Printing Conference at Columbia University, May 19-22, 1982. *New York: School of Library Science, Columbia University, 1983.*

Solomon, Richard. Chuck Close: Handmade Paper Editions. *New York: Pace Editions, Inc., 1983.*

Toale, Bernard. The Art of Papermaking. *Worcester, MA: Davis Publications, 1983; 22-87, 114, 116.*

Turner, Silvie. Paper as Image. *London: Arts Council of Great Britain, 1983.*

Turner, Silvie and Birgit Skiöld. Handmade Paper Today. *London: Lund Humphries Publishers, 1983; 31, 35, 98, 104-105.*

Brouwer, Bert. Paper Transformed: A National Exhibition of Paper Art. *Terre Haute, IN: Turman Gallery, Indiana State University, 1984.*

Wong, Paul. Paper as Paint. *New York: Fashion Institute of Technology, 1985.*

Siegenthaler, Fred. Strange Papers. *1986.*

Toale, Bernard. Large Scale Works in Paper. *Boston: Bank of Boston, 1986.*

Dieu Donné Papermill: 15 Years, February 1 – March 9, 1991. *New York: Book Arts Gallery, The Center for Book Arts, 1991.*

Farmer, Jane. "Contemporary Hand Papermaking in the United States." *In* International Biennial of Paper Art: Paper and Nature. *Düren, West Germany: Leopold-Hoesch Museum, Wienand Verlag, 1992; 54. (First printed in International Conference of Hand*

Papermakers. Boston: Carriage House Press, 1981.)

Stein, Donna. Vincent FitzGerald & Company: A Catalogue Raisonné of the First 26 Books Published by Vincent FitzGerald & Company from 1981 – 1992. *New York: Vincent FitzGerald & Co., 1992.*

Gosin, Susan. Statement in Made in the Midwest: Walter Hamady's 6451 Students. *Racine, WI: Charles A. Wustum Museum of Fine Arts, 1993.*

Paper – The Essential Material: Selected Book Projects at Dieu Donné Papermill, January 7-29, 1993. *New York: Harper Collins, 1993. Introduction by Mina Takahashi.*

Takahashi, Mina and Paul Wong. "Working the Mill: Selected Artist Collaborations at Dieu Donné Papermill." *In* Art at FGIC. *New York: Financial Guaranty Insurance Company, 1993.*

Crivelli, Ricardo A. Notas Sobre Papel Hecho a Mano. *Buenos Aires: Taller GRABART, 1994; 171.*

Dawson, Sophie and Silvie Turner. A Hand Papermaker's Sourcebook. *New York: Design Books, 1995; 28, 47, 99, 105, 110.*

Hansen, Trudy. Printmaking in America: Collaborative Prints and Presses, 1960-1990. *New York: Harry N. Abrams, 1995; 51.*

Collischan, Judy. Paper Spaces. *Purchase, NY: Neuberger Art Museum, State University of New York, 1997.*

Illuminations: Bookworks of Susan Weil. *Stockholm: National Museum, 1997; 24, 36-37, 44-47.*

Ishac, Allan. New York's 50 Best Places to Take Children. *New York: City & Company, 1997; 32-33.*

Hiebert, Helen. Papermaking with Plants. *Pownal, VT: Storey Books, 1998; 67, 103.*

Morton, Robert, ed. Sandy Skoglund: Reality Under Siege: A Retrospective. *New York: Harry N. Abrams, 1998; 33, 45, 47, 65-66, 175, 193, 199-231, and sample paper insert.*

Stein, Donna. Susan Weil: Full Circle *(Black Mountain Dossier, no. 2). Black Mountain, NC: Black Mountain College Museum and Arts Center, 1998.*

Turner, Silvie. The Book of Fine Paper. *New York: Thames and Hudson, 1998.*

Nagy, Peter. "Symbols and the Container for Symbols." Mirror: Arlene Shechet. *New York: Elizabeth Harris Gallery, 1999.*

Paper Trail: Douglass Howell and Four Pioneers in American Handmade Paper: Laurence Barker, Walter Hamady, Clinton Hill, Golda Lewis. *New Brunswick, NJ: Rutgers Center for Innovative Print and Paper, 2000.*

Themes & Variations: The Publications of Vincent FitzGerald & Co., 1980 – 2000. *New York: Columbia University and Vincent FitzGerald & Co., 2000; 84, 86.*

Performance Drawings *(The Drawing Center's Drawing Papers, no. 20). New York: The Drawing Center, 2001.*

Articles and Reviews

"Rags to Riches." New York Magazine *(24 October 1977): 73.*
Schiller, Marlene. "The American Community of Hand Papermakers." American Artist *(August 1977): 42.*

Markell, Eliot. "Pulp Art." Art Workers News *(February/March 1978): 13.*

Clark, Cynthia. "An Industry Survey: Papermaking Revisited." Print Collector's Newsletter *(July/August 1979): 81.*

"New Editions: Chuck Close." Art News *(April 1982): 102.*

Toale, Bernard. "Papermaking in the USA." Fiberarts *(July/August 1982): 34-38.*

Perrault, John. "Paperworks." American Craft *(August/September 1982): 2-7.*

"The Handmade Paper Book." American Craft *(August/September 1982): 8-10.*

Tully, Judd. "Paper Chase." Portfolio *(May/June 1983): 78-85.*

Cohen, Ronny. "Paper Routes." Art News *(October 1983): 79-80, 83-85.*

Coffin, Anne. "Creative Collaborations: Artists & Printmakers Part II." Downtown *(November 1983).*

Scarborough, Jessica. "Sculptural Paper: Foundations and Directions." Fiberarts *(March/April 1984): 32-35, 73.*

Bruckner, D. J. R. "With Art and Craftsmanship, Books Regain Former Glory." New York Times Magazine *(28 October 1984): 40, 49.*

Baxter, Barbara A. "Creative Papermaking at Firestone Library." Town Topics *(Princeton, 11 January 1989): 3.*
Sinclaire, Estelle. "Princeton's Firestone Library Showing Rich and Diverse Styles of this Fine Art." Time Off *(Princeton, 18 January 1989).*

Henry, Gerrit. "Vincent FitzGerald: Five Ruminations." Print Collector's Newsletter *(March/April 1990): 10-12.*

Silverberg, Robbin Ami. "Tim Barrett Kicks Off Dieu Donné Lectu-e Series at Granary Books." Book Arts Review *(spring 1990): 3, 6.*

Mathews-Berenson Margaret. "A Resource for Handmade Papers." American Artist *(November 1991): 51-62.*

"Lintas: Worldwide - An Advertising Agency with an Eye on Art" Corporate Art News *(December 1991).*

Dubansky, Mindell. "Fifteen Years at Dieu Donné Papermill." Hand Papermaking *(winter 1991): 15-18.*

Takahashi, Mina. "Azure Paper from Dieu Donné." Hand Papermaking *(winter 1991): 19.*

Fox, Margalit. "Papermill: Art for the Sake of Artists." New York Newsday *(2 January 1992): 42, 61.*

Evans, Poppy. "Carter Hodgkin: Making Photoimaged Paper." Screen Printing *(June 1992): 109.*

Casey, Shawn. "Cast in Paper." Sky *(September 1992): 88-97.*

Meier-Husby, Barbara, Jesse Munn, Terry Wallis, and Mary Wooton. "Quality Materials: Handmade Paper. Taking a Closer Look." *In* The Book and Paper Group Annual *(Stanford: American Institute of Conservation, 1993). 61-65.*

Leimbach, Dulcie. "Papermaking." New York Times *(11 February 1994).*

"Museum Exhibiting Books Enhanced by Special Paper." Berkshire Eagle *(13 August 1994): 33*

"A Move to Help Make More Paper the Old-Fashioned Way: By Hand." New York Times *(13 November 1994), The City section.*

Sult, Amy. "Two exhibits offer unusual expressions in papermaking." Kalamazoo Gazette *(13 November 1994).*

"Learning About Paper from the Ground Up." Tribeca Trib *(December 1994): 25.*

Cohen, Esther. "Making Paper for Art's Sake." New York Newsday *(16 March 1995): City Neighborhoods, B7.*

Brower, Steven. "Autoparticulates." Hand Papermaking *(summer 1995): 2-4.*

Doherty, Kathleen. "Paper in New York." Hand Papermaking *(summer 1995): 30-33.*

"Curtains: Pulp Fashion." Fiberarts *(September/October 1995): 88.*

Gosin, Susan. "An Interview with Bart Wasserman." Hand Papermaking *(winter 1995): 5-9.*

Williams, Stan. "Behind The Scene: Paper View." DNR *(15 December 1995).*

"Gobbene, by Grimanesa Amoros." American Craft *(February/March 1996).*

Liebenson, Bess. "Focus on an Antique Medium Called Paper." New York Times *(10 March 1996): 19.*

"Sales-Dieu Donné." Time Out New York *(20-27 March 1996).*

Cotter, Holland. "Byron Kim." New York Times *(5 July 1996), Art in Review section.*

Cotter, Holland. "Sculpture that Basks in Summer." New York Times *(9 August 1996): C-24.*

"Ken Polinskie/Dieu Donné Papermill." American Craft *(December 1996/ January 1997): 70.*

"Dieu Donné Papermill." Time Out New York *(9-16 January 1997).*

Glueck, Grace. "Christy Rupp: Moisture Seekers." New York Times *(10 January 1997): A-22.*

"Dieu Donné Papermill: Symposium." American Craft *(February/March, 1997).*

"Art of the Matter: Seven Collaborations in Paper Art." New Times *(Tempe, AZ, 21 August 1997): Calendar, 38.*

Rausch, Gary. "Putting It All on Paper." Tribune *(Tempe, AZ, 28 August 1997): 25.*

MacLaren, Emily. "Pulp Fiction." Surface, *no. 14, 1998.*

"Papermaking." *In* Bridges Elementary School Yearbook. *New York: Bridges Elementary School, 1998); 22.*

"Nancy Cohen: Only Connect." Sculpture *(January 1998).*

"Arlene Shechet." On Paper, The Journal of Prints, Drawings, and Photography *(January/February 1998).*

"Paper +." NY Arts *(May 1998): 43.*

Berstein, Jennifer. "Joe Wilfer: Collaborations in Paper and Printmaking." Hand Papermaking *(summer 1998): 7-9.*

Hirsch, Faye. "Paper +: Works on Dieu Donné Paper." On Paper, The Journal of Prints, Drawings, and Photography *(July/August 1998): 49.*

Egan, Maura. "Arts and Crafts: Handmade Dreams." New York *(3 August 1998): 44.*

"Emiko Kasahara." Art On Paper *(November/December 1998): 53.*

Basker, J. Taylor. "Dieu Donné Papermill, Inc." Westbetharts.com *(25 March 1999).*

"Elena del Rivero." Boletín Cultural 219-21 *Consulado General de España, New York, June/August 1999.*

Johnson, Ken. "Drawn to Scale." New York Times *(6 August 1999): Art in Review section.*

"Byron Kim, Blue Night." Art On Paper *(September/October 1999).*

Deery, Gail and Mina Takahashi. "Economic Development in Ecuador." Hand Papermaking *(winter 1999): 20-26.*

Armada, Alfonso. "Elena del Rivero expone en Nueva York sus últimas obsesiones." El Mundo *(7 December 1999).*

Monaghan, Peter. "Notes from Academe: Ecuador." The Chronicle of Higher Education *(10 March 2000): B 2.*

Gardner, Paul. "New Nature: Recent Drawings by Richmond Burton." Art On Paper *(March/April 2000): 48.*

"Mel Bochner. Language Is Not Transparent." Art On Paper *(March/April 2000).*

DeNitto, Emily. "Groups Need More Elbow Room." Crain's New York Business *(27 March – 2 April 2000).*

"Be Still My Heart: Winifred Lutz Waxes Poetic in Her Sculpture Show '(stillness)'." Philadelphia Weekly *(5 April 2000): 51.*

Rice, Robin. "A Thread Runs Through It." Philadelphia City Paper *(6 April 2000).*

"Dieu Donné Papermill To Present Art Award." Walden's Paper Report *(10 April 2000).*

Garcia-Fenech, Giovanni. "Dieu Donné Art Awards." Artnet News, artnet.com *(4 May 2000).*

Van Dongen, Susan. "Pulp Faction." Princeton Packet *(30 June 2000): 19.*

"Paper People." Art On Paper *(July/August 2000): 16-17.*

Bischoff, Dan. "The Flax of Life." Star-Ledger *(4 August 2000): Art, 11.*

"Paper, Ubiquitous & Artful." U.S.1. *(9 August 2000).*

Genzlinger, Neil. "Artists Make Traditional Canvas into the Medium." New York Times *(27 August 2000).*

"People & Places." American Craft *(August/September 2000).*

"New Collaboration with Juan Sánchez." Printshop Newsletter, no. 4 *Lower East Side Printshop, New York, September 2000.*

Durgin, Michael. "Book Review: Dieu Donné's Hand Papermaker's Cookbook." Hand Papermaking Newsletter, no. 52 *(October 2000).*

"Talent - Alan Shields." New York *(2 October 2000).*

Altchek, Ruth. "In the Play Room." Art On Paper *(November/December 2000): 30.*

"Donald Baechler, York House Suite." Art On Paper *(November/December 2000): 79.*

La Rocca, Claudia. "Paper Trail." Art On Paper *(November/December 2000): 89.*

"LandEscape." Goings On, circline.com *(1 November 2000).*

Doherty, Kathleen. "Paper Trail." Hand Papermaking *(winter 2000): 46-47.*

"Euraba Paper Dreaming." Hand Papermaking *(winter 2000): 24.*

Wellish, Marjorie. "Alan Shields." Art News *(December 2000): 156.*

Goodman, Jonathan. "Alan Shields at Paula Cooper and Nicholas Davies." Art in America *(January 2001): 117.*

"Out of the Floating Waves." Asian American Press *(5 January 2001): 12.*

Dube, Ilene. "Paper Maker Embeds Natural Elements into Medium." Greater Princeton Extra *(12 January 2001): 1.*

Maksymowicz, Virginia. "High, Low, and In-Between." Sculpture *(March 2001): 26-29.*

"Diana Cooper." Art On Paper *(March/April 2001): 76.*

"Feb 25: The New York Public Library." New York Times *(11 March 2001).*

Frank, Douglas. "Portrait of the Artist: The Works of Mel Edwards, Sculptor and Printmaker." Rutgers Focus *(16 March 2001): 4-5.*

"Drawn to Read." Art On Paper *(May/June 2001): 29-30.*

Ruyak, Jacqueline. "In Print: The Dieu Donné Hand Papermakers' 'Cookbook'." Surface Design Journal *(Spring 2001): 58.*

DeNitto, Emily. "Tough Times Ahead, Depending on Whom You Ask." Crain's New York Business *(21-27 May 2000): 65.*

Plagens, Peter. "Our Man in Venice." Newsweek *(4 June 2001): 54-55.*

Glossary of Papermaking Terms

ABACA: Also known as Manila hemp (comes from the leaves of *Musa textilis*, a variety of banana plant grown in Philippines) that can produce lustrous, fairly translucent paper.

ACID-FREE: Paper which has a pH level above 6.5 or is free from any acid content or other substance likely to prove detrimental to the paper's permanence.

ALKALI: A caustic substance (for example lye, lime, wood ash, or soda ash) used when cooking plant fibers to remove gums, waxes, starch, and other noncellulosic materials.

ALUM: A complex salt, commonly aluminum sulfate, added as a hardener with sizing or a mordant when dyeing pulp. If used in excessive quantities, its high acidity will cause irreversible damage to the paper.

ARCHIVAL: A paper with long-lasting qualities, with a pH value between 5.5 and 8, lignin-free, or buffered with an alkaline reserve; also usually lightfast.

ARTCOR: A waterproof material used for shaped deckles and stencils. The sheets, which are styrofoam sandwiched between two layers of plastic, cuts easily with an X-acto knife, and can be taped temporarily to existing deckles.

BAGASSE: Sugar cane fiber, a tall perennial grass.

BAST FIBER: Fiber derived from the inner bark of woody plants such as mitsumata, gampi, and kozo and from certain herbaceous plants such as flax, jute, hemp, and ramie.

BEATER: Also known as Hollander beater, a seventeenth-century Dutch invention that is still in use today to macerate fibers into pulp by cutting or crushing fibers between the rotating bars of the beater roll and the stationary bars of the bed plate. The term "Reina beater" refers to a beater fabricated by David Reina, Brooklyn, NY. The term "Valley beater" refers to a test-beater used by the machinemade paper industry.

BEATER ROLL: Barred roll used to crush, cut, and fibrillate the fiber as it passes across the bed plate of the beater.

BEATING: The physical or mechanical process by which fibers are cut, shredded, macerated, and separated, in water, until they form a usable pulp. Common beating mechanisms include Hollander beaters, whiz mixers, blenders, and mallets. (See Hydration, Fibrillation, Hydrogen Bonding.)

BED PLATE: A set of bars or blades, sitting underneath and in close proximity to the beater roll. The bed plate and roller function together to macerate fibers.

BLEEDING: When pigments and dyes fail to adhere to the fibers they are meant to color, or when spent dyes are not fully rinsed.

BUFFER: An alkaline substance (usually calcium carbonate or magnesium carbonate) that helps protect paper from acidity in the paper and the environment.

BRITTLENESS: The quality in a paper of being hard and susceptible to breaking when bent or folded.

CALCIUM CARBONATE: Ground powder of pure limestone used for adding an alkaline reserve in paper to increase longevity of paper and enhance opacity. (See Buffer.)

CALENDER: A physical process by which the surface of the paper is made smooth or glazed. While referring to a process of passing paper between hard or polished rollers, a smooth surface (on one side of the paper) can be achieved by metal-drying.

CASTING: Making relief forms or sculpture in-the-round with paper pulp or newly-formed sheets of paper. (See Laminate Casting and Pulp Casting.)

CELLULOSE: Chemically, a high-molecular-weight polymer (or chain) of glucose; the chief component of plant tissue. The attraction between cellulose molecules is the principal source of fiber-to-fiber bonding. Cotton and linen fibers contain the most generous amounts of cellulose and are ideal for papermaking.

CHARGE: Add pulp to a vat to replenish vat stock: "charge" a vat.

COTTON RAG HALF-STUFF: Cotton scraps from the garment industry that is partially processed and in near-thread form. Sold as a chunky mass, sometimes moist.

CLOUDY: When a sheet is formed irregularly due to fast draining pulp or poor formation.

COCKLING: Surface undulation or wavy edges due to uneven drying.

COTTON LINTERS: The coarser, shorter fibers left on the cotton seed after the long staple fiber has been ginned away for use in the textile industry. They are cut from the seed, cooked, and formed into pulp sheets. Cotton linters are low-shrinkage fibers and recommended for hand casting. Because of the short fibers, they are not a good basis for book papers, which need longer fibers for durability against handling and folding.

COUCHING: Transferring a freshly made sheet of paper from the mould surface onto a dampened felt. Rhymes with "smooching"; from the French word "coucher," meaning to lie down or put to bed.

DANDY ROLL: A large cylindrical roll on the finishing end of a Fourdrinier papermaking machine, which imparts a pattern on or impresses a watermark into the newly formed paper.

DECKLE: The removable frame which fits onto the mould to contain pulp and determine the size and shape of the sheet.

DECKLE BOX: A deckle which has deep sides (like a box) that clamps onto a mould. The deckle box is partially submerged in a vat of water, pulp is poured into it, then the box is lifted out of the water.

DECKLE EDGE: The natural, feathery edge of a handmade sheet of paper, created by the deckle.

DRYING: Paper can be dried in various ways, including air drying, restraint drying by applying pressure on top of the paper, or by brushing or pressing paper onto a surface made of metal, glass, wood, or other material.

DYES: Water-soluble coloring agents which chemically penetrate structure of fiber. Dyes, whether synthetic or natural (i.e., onion skin, walnut husks, insects), are not very lightfast when compared to pigments and are susceptible to fading due to exposure to UV light. (See Pigment.)

EMBEDDING: Incorporating objects or materials in a sheet of paper, either by sandwiching them between sheets or layering pulp over the object to hold the material in place.

EMBOSSING: Creating a raised or depressed surface design by laying a material onto a wet sheet of paper and pressing it. Embossing can be done at the initial pressing or later, by dampening a dry sheet.

FELT: The material (traditionally a woolen blanket) onto which a newly formed sheet of paper is couched. The paper stays on the felt until it has been pressed, then is removed and placed in a drying system.

FIBER: Cellulose-based material derived from plant matter which forms the basis of a sheet of paper.

FIBRILLATION: The process of breaking up the surface of individual cellulose fibers during beating, thereby creating tiny hairs that encourage bonding.

FLAX: The inner bark of *Linum usitatissimum* used to make linen cloth.

FLOCCULATE: To affect the pulp so that fibers clump together. (See PNS, under Formation Aid).

FORCED-AIR STACK DRYER: A system in which newly formed-and-pressed paper is stacked with interleaving material, put under weight, and dried by forcing air through the stack. Can be fashioned from blotters, corrugated cardboard, and a box-fan.

FORMATION: In papermaking, the fiber distribution throughout a sheet observed by holding the sheet up to light.

FORMATION AID: A substance introduced into the vat stock that improves distribution of fibers by decreasing fiber-to-fiber flocculation and slowing drainage during sheet-forming. Natural formation aids include the sticky extract of the tororo-aoi plant (which is grown in Japan) and okra. Synthetic formation aids that are available through papermaking suppliers include PMP, PNS and PEO. PMP and PEO are used with sized or pigmented pulps as they are cationic (+) agents that do not react as a flocculent in the presence of sizing, retention aids, and other anionic (-) agents. PNS can only be used with unsized or unpigmented pulps; it acts as a flocculent with sized and pigmented pulps. (See Retention Aid.)

GAMPI: The rarest of the three plants used in Japan for papermaking. It is harvested wild and cannot be cultivated and yields a highly lustrous and insect-repellent paper. Fiber length is between 2 and 4.5 mm. Philippine gampi is a common term for "salago," a fiber cultivated in the Philippines with similar aesthetic properties as Japanese gampi.

GELATIN: A glutinous material, extracted from animal tissue by boiling, that is used to surface-size a sheet of paper, making the paper less water-absorbent.

GRAIN: The directional alignment of fibers in a sheet of paper. Generally, a handmade sheet has a less pronounced grain than a machine made paper.

HIGH-SHRINKAGE PULP: Pulp that is fibrillated to an extreme degree so as to cause the surface area of a sheet made from pulp to shrink substantially when dried. For example, linen rags beaten for 8 hours will shrink 10% when dried under weight and about 50% when air-dried.

HYDRATION: A process that occurs during beating in which the bruised fibers begin to absorb water more readily.

HYDRAULIC PRESS: A platen press which applies pressure vertically (as opposed to rolling) by means of a hydraulic system.

HYDROCAL: A finer grade of gypsum that sets up more quickly and is denser and harder than pottery plaster. Can be used for making moulds for paper casting.

HYDROGEN BONDING: The electrostatic attraction between hydrogen atoms in water helps to bring fibers together as the sheet is formed. As the final molecules of water leave the sheet, hydrogen bonds form between the fibers.

"KISS OFF": When a sheet of paper which is still on the mould is unsatisfactory, the mould surface is touched onto the water surface of the vat, causing the pulp to fall back into the vat.

KOZO: Paper mulberry plant originating in Japan and Asia. The inner bast fiber is strong and long with a fiber length between 3 and 25 mm. More than one-half of Japan's paper is made of kozo.

LAID LINES: The closely spaced lines seen in paper which are formed by the wires of a laid mould.

LAID MOULD: A mould whose cover is made of closely spaced, parallel wires held in place by more widely spaced perpendicular chain wires sewn to support ribbing.

LAMINATE CASTING: Applying already formed and lightly pressed paper onto an armature or a form.

LAMINATING: Combining layers of paper by couching one newly formed sheet on top of another to create a single sheet.

LIGHT FASTNESS: Resistance of color to fading or changing.

LIGNIN: The fiber-bonding material found mainly in woody plants. It resists hydrogen bonding of fibers, therefore it must be removed, usually by cooking, before the fiber can be used for papermaking.

LYE: Sodium hydroxide (NaOH). A caustic, harsher than soda ash, used to break down tough plant fibers for papermaking.

MACERATE: Bruising and separating of the fibers during the beating process

METHYL CELLULOSE: A powdered substance that is mixed with water and is pH neutral and archival and has many uses. It is usually used as an adhesive and can be added to the wet pulp to promote fiber-to-fiber bonding and strengthen paper castings; it can be brushed onto dry paper as a surface size, or added to colored pulp to improve the pulp's flowing consistency for pulp painting.

METAL-DRYING: Pressing and drying newly formed-and-pressed paper on metal sheets, thus causing it to take on the slick surface of the metal.

MITSUMATA: A plant used in Japan whose inner bark yields a lustrous and warm-toned paper. Fiber length is between 1 and 5 mm.

MORDANT: A chemical used to cause increased fixation of the dye to the fiber.

MOULD: A rectangular wooden frame covered with screen or laid wire upon which paper is formed and drained.

MOULD-MADE PAPER: Paper made on a cylinder-mould machine. It is generally distinguished by having two "deckle edges." Formed by wax strips on the cylinder that repel fibers and create a feathery edge between sheets on the cylinder.

NO-SEE-UM NETTING: Finely woven polyester mesh that is finer than mosquito netting. Used as a screen mesh for Japanese sugetas, lining colanders for draining pulp, and creating removable watermark stencils.

OVERBEATING: A commonly used but somewhat inaccurate term to describe extreme fibrillation of fiber to achieve maximum pigmentation, high-shrinkage, and/or translucency of the pulp.

PELLON: Polyester sheet material sold in fabric stores as interfacing and available in varying thicknesses. Can be used as a substitute for papermaking felts.

POST: A stack of newly formed sheets of paper between felts.

PH: The pH value measures acidity or alkalinity of the paper. Values of pH range from 0-14, with 7 indicating neutrality, numbers less than 7 indicating levels of acidity, and numbers greater than 7 indicating levels of alkalinity.

PIGMENT: Coloring matter in the form of insoluble, finely ground particles, which mechanically deposits to the outside of the fibers they are coloring. Generally, a retention agent is needed to adhere the pigment to the fiber.

PRESSING: Compaction of freshly formed sheets to remove excess water and compress the fibers sufficiently so that the sheet can be lifted from the felt without falling apart.

PULP: The aqueous mixture of macerated fibrous material from which paper is made.

PULP CASTING: Laying dense, wet pulp into a mold or onto a surface and pressing and extracting excess water with a sponge until the pulp is firmly compressed to the surface. When dry, the pulp is removed and takes the form of the surface upon which it was cast.

PULP SPRAYING: Spraying pulp onto a mold, armature, or a surface through an air-compressor-driven spray gun or pattern-pistol.

PVA: Polyvinyl acetate is an archival adhesive that is semireversible and dries clear.

RAG: Woven fabric used for papermaking. The term "rag" or "all-rag" properly describes a sheet made entirely from woven fabric. Today however, the term is sometimes used as a misnomer to describe paper made from cotton linters.

RICE PAPER: A misnomer for Asian papers, which are actually made from the bast fiber of plants, most often from kozo, the paper mulberry plant. True "rice paper" is a sheet cut spirally from the inner pith of the rice paper plant (*tetrapanax papyriferus*) very much like wood veneer.

RETENTION AID: A substance that changes the ionic charge of the pulp from anionic (-) to cationic (+) to attract it to additives which are generally anionic. It is added to pulp to attract pigments to the fiber to retain color and to prevent bleeding.

RETTING: Rotting down fiber source (whether plant or rags) by soaking in alkaline solution or water in order to break down the fiber to the usable cellulose material. Often, lime, or calcium hydroxide ($Ca[OH]_2$), is used as a retting alkali.

RIBS: Thin wooden struts fixed into the frame of the mould to support the wire mould cover.

SHAKE: The action of the papermaker when s/he dips a layer of pulp onto the mould and moves the mould in a front-to-back and side-to-side motion to allow the watery film of pulp to settle in an even deposit.

SIZING: Any material added to the paper that increases the paper's resistance to penetration by liquids. "Internal" size refers to a sizing agent added to pulp prior to sheet formation. A sizing agent applied after a sheet is formed and dried is called an "external" or "surface" size.

SODIUM CARBONATE: Otherwise known as soda ash (Na_2CO_3), a mild caustic used to cook fiber to remove noncellulosics. (See Alkali.)

SPUR: A group of sheets air-dried together rather than individually.

STOCK: Another term for pulp.

SUGETA: A Japanese papermaking mould that has a flexible removable screen made of bamboo splints woven together ("su") and a hinged deckle and support frame ("geta").

VAT: A tub or vessel that holds pulp for sheet-forming.

WASHI: From the Japanese "wa," meaning Japan, and "shi," meaning paper. "Washi" refers to Japanese handmade paper of any type.

WATERLEAF: Paper that has no sizing agent and is very absorbent.

WATERMARK: A translucent design or image in a sheet of paper, formed by varying the density of the pulp at certain points in the sheet by raising or lowering the surface of the mould at selected places. Traditionally the design was fashioned from wire and sewn onto the screen surface of the mould. Contemporary methods include watermarks fashioned from vinyl signage material, self-adhesive contact shelving paper, or tape, which are easily applied and removed from the screen.

WHIZ MIXER: A device with a mixing attachment that hydrates pulp sheets or already processed pulps. It is not mechanically designed to beat rags into pulp.

WIRE SIDE: The side of the sheet which faces the surface of a mould.

WOVE: A mould surface made of wire or plastic mesh that imparts a more anonymous surface than a laid mould.

Definitions for the Papermaking Glossary have been cross-referenced with the following sources:

Barrett, Timothy. Japanese Papermaking. *New York: Weatherhill Inc., 1992.*

Bower, Peter. Turner's Later Papers. *London: Tate Gallery Publishing, 1999.*

Turner, Silvie. Which Paper? *New York: Design Books, 1994.*

Turner, Silvie. The Book of Fine Paper. *London: Thames and Hudson, 1998.*

Zmolek Smith, Gloria. Teaching Hand Papermaking. *Cedar Rapids: Z Paperpress, 1989.*

Carriage House Paper Supplies Catalog

Lee S. McDonald, Inc. Equipment and Supplies Catalog

Twinrocker Papermaking Supplies Catalog

Index of Names

AS THEY APPEAR IN ESSAYS

Photography Credits

WORKS OF ART

Jack Abraham: *fig. 11;* John Back: *figs. 42, 50;* Laurence Barker: *fig. 3;* D. James Dee: *figs. 1, 26, 44, 52, 54, 57;* Liz Deschenes: *fig. 24;* Ming Fay: *fig. 68;* courtesy of Gemini G.E.L.: *fig. 6;* Arnold Helbling: *fig. 59;* Hiro Ihara Studio: *fig. 37;* Richard Johnson: *fig. 43;* Margaret Lanzetta: *fig. 47;* courtesy of Golda Lewis: *fig. 58;* courtesy of Winifred Lutz: *fig. 5;* Bill Milne: *figs. 17, 18;* Wayne McCall: *fig. 32;* Peter Muscato: *fig. 31;* Bill Orcutt: *fig. 62;* courtesy of Pace Editions, Inc.: *fig. 8;* Ken Pelka: *fig. 53;* Benjamin Shiff: *fig. 69;* Sandy Skoglund: *fig. 67;* Alexandra Soteriou: *fig. 2;* Larry Stanton: *fig. 7;* Michelle Stuart: *fig. 9;* Jeff Sturges: *fig. 51;* Mina Takahashi: *fig. 64;* Taylor Photo: *figs. 4, 27, 28, 56;* Jean Vong: *figs. 12, 13, 14, 15, 16, 19, 20, 21, 22, 23, 25, 29, 30, 34, 35, 36, 39, 40, 41, 46, 48, 49, 60, 63, 65, 66;* Tom Warren: *fig. 45;* Paul Wong: *figs. 55, 61;* Zindman/Fremont, NY: *figs. 10, 38;* Carson Zullinger: *fig. 33*

DOCUMENTARY PHOTOGRAPHS AND PORTRAITS

Alan Ackoff: *42;* John Back: *27 bottom;* Geoffrey Clifford/Black Star: *24;* Gail Deery: *32 top;* courtesy of Dieu Donné Papermill: *26, 27 top, 30 top, 30 bottom, 31 middle, 31 bottom, 33;* Matthias Geiger: *34 bottom;* Susan Gosin: *25;* Stacey Lynch: *28;* Bill Milne: *11, 40, 41;* Mina Takahashi: *31 top, 32 middle, 32 bottom, 34 top;* Christopher Ytuarte: *30 middle*

Rags to Riches was designed and typeset by Russell Maret.
The typefaces are Quadraat and Scala Sans.

Sue Henger copyedited and indexed the manuscript.

3,500 copies were printed and bound by Friesens.
The text paper is Luna Matte, an acid-free paper.

Each copy includes a bookmark handmade at Dieu Donné Papermill.